A Review of Algebra, by Romeyn Henry Rivenburg

The Project Gutenberg EBook of A Review of Algebra, by Romeyn Henry Rivenburg This eBook is for the use of anyone anywhere at no cost and with almost no restrictions whatsoever. You may copy it, give it away or re-use it under the terms of the Project Gutenberg License included with this eBook or online at www.gutenberg.org

Title: A Review of Algebra

Author: Romeyn Henry Rivenburg

Release Date: January 9, 2012 [EBook #38536]

Language: English

Character set encoding: ISO-8859-1

*** START OF THIS PROJECT GUTENBERG EBOOK A REVIEW OF ALGEBRA ***

[Transcriber's Note:

This book includes extensive mathematical expressions and equations, which can not always be easily represented in plain text. The reader is encouraged to download the HTML version of the text, which represents the math more clearly.

For the plain text version, the following conventions are used:

Mixed fractions are represented by a dash with no spaces, while subtraction is represented by a dash with spaces on either side. For example: 1-1/2 is "one and one half." 1 - 1/2 is "one minus one half."

The "sideways-8" symbol for infinity is represented as [infinity].

Square, cube, and other roots are shown by raising a quantity to the appropriate fractional power. For example: [4]^(1/2) is "the square root of 4." [x]^(1/n) is "the nth root of x."

Extra parentheses have been added as needed to clarify the correct order of operations.]

A REVIEW OF ALGEBRA

BY ROMEYN HENRY RIVENBURG, A.M.

HEAD OF THE DEPARTMENT OF MATHEMATICS THE PEDDIE INSTITUTE, HIGHTSTOWN, N.J.

[Illustration]

AMERICAN BOOK COMPANY NEW YORK CINCINNATI CHICAGO

COPYRIGHT, 1914, BY ROMEYN H. RIVENBURG.

COPYRIGHT, 1914, IN GREAT BRITAIN.

A REVIEW OF ALGEBRA.

E. P. 6

PREFACE

In most high schools the course in Elementary Algebra is finished by the end of the second year. By the senior year, most students have forgotten many of the principles, and a thorough review is necessary in order to prepare college candidates for the entrance examinations and for effective work in the freshman year in college. Recognizing this need, many schools are devoting at least two periods a week for part of the senior year to a review of algebra.

For such a review the regular textbook is inadequate. From an embarrassment of riches the teacher finds it laborious to select the proper examples, while the student wastes time in searching for scattered assignments. The object of this book is to conserve the time and effort of both teacher and student, by providing a thorough and effective review that can readily be completed, if need be, in two periods a week for a half year.

Each student is expected to use his regular textbook in algebra for reference, as he would use a dictionary,--to recall a definition, a rule, or a process that he has forgotten. He should be encouraged to *think* his way out wherever possible, however, and to refer to the textbook only when *forced* to do so as a last resort.

The definitions given in the General Outline should be reviewed as occasion arises for their use. The whole Outline can be profitably employed for rapid class reviews, by covering the part of the Outline that indicates the answer, the method, the example, or the formula, as the case may be.

The whole scheme of the book is ordinarily to have a page of problems represent a day's work. This, of course, does not apply to the Outlines or the few pages of theory, which can be covered more rapidly. By this plan, making only a part of the omissions indicated in the next paragraph, the essentials of the algebra can be readily covered, if need be, in from thirty to thirty-two lessons, thus leaving time for tests, even if only eighteen weeks, of two periods each, are allotted to the course.

If a brief course is desired, the Miscellaneous Examples (pp. 31 to 35, 50 to 52), many of the problems at the end of the book, and the College Entrance Examinations may be omitted without marring the continuity or the comprehensiveness of the review.

ROMEYN H. RIVENBURG.

CONTENTS

PAGES

OUTLINE OF ELEMENTARY AND INTERMEDIATE ALGEBRA 7-13

ORDER OF OPERATIONS, EVALUATION, PARENTHESES 14

SPECIAL RULES OF MULTIPLICATION AND DIVISION 15

CASES IN FACTORING 16, 17

FACTORING 18

HIGHEST COMMON FACTOR AND LOWEST COMMON MULTIPLE 19

FRACTIONS 20

COMPLEX FRACTIONS AND FRACTIONAL EQUATIONS 21, 22

SIMULTANEOUS EQUATIONS AND INVOLUTION 23, 24

SQUARE ROOT 25

THEORY OF EXPONENTS 26-28

RADICALS 29, 30

MISCELLANEOUS EXAMPLES, ALGEBRA TO QUADRATICS 31-35

QUADRATIC EQUATIONS 36, 37

THE THEORY OF QUADRATIC EQUATIONS 38-41

OUTLINE OF SIMULTANEOUS QUADRATICS 42, 43

SIMULTANEOUS QUADRATICS 44

RATIO AND PROPORTION 45, 46

ARITHMETICAL PROGRESSION 47

GEOMETRICAL PROGRESSION 48

THE BINOMIAL THEOREM 49

MISCELLANEOUS EXAMPLES, QUADRATICS AND BEYOND 50-52

PROBLEMS--LINEAR EQUATIONS, SIMULTANEOUS EQUATIONS, QUADRATIC EQUATIONS, SIMULTANEOUS QUADRATICS 53-57

COLLEGE ENTRANCE EXAMINATIONS 58-80

OUTLINE OF ELEMENTARY AND INTERMEDIATE ALGEBRA

~Important Definitions~

Factors; coefficient; exponent; power; base; term; algebraic sum; similar terms; degree; homogeneous expression; linear equation; root of an equation; root of an expression; identity; conditional equation; prime quantity; highest common factor (H. C. F.); lowest common multiple (L. C. M.); involution; evolution; imaginary number; real number; rational; similar radicals; binomial surd; pure quadratic equation; affected quadratic equation; equation in the quadratic form; simultaneous linear equations; simultaneous quadratic equations; discriminant; symmetrical expression; ratio; proportion; fourth proportional; third proportional; mean proportional; arithmetic progression; geometric progression; S [infinity]

~Special Rules for Multiplication and Division~

1. Square of the sum of two quantities. $(x + y)^2$.

2. Square of the difference of two quantities. $(x - y)^2$.

3. Product of the sum and difference of two quantities. $(s + t)(s - t)$.

4. Product of two binomials having a common term. $(x + r)(x + m)$.

5. Product of two binomials whose corresponding terms are similar. $(3x + 2t)(2x - 5t)$.

6. Square of a polynomial. $(m - n/3 + k)^2$.

7. Sum of two cubes. $(x^3 + y^3)/(x + y) = x^2 - xy + y^2$.

8. Difference of two cubes. $(x^3 - y^3)/(x - y) = x^2 + xy + y^2$.

9. Sum or difference of two like powers. $(x^7 + y^7)/(x + y)$, $(x^5 - y^5)/(x - y)$, $(x^4 - y^4)/(x - y)$, $(x^4 - y^4)/(x + y)$.

~Cases in Factoring~

1. Common monomial factor. $mx + my - mz = m(x + y - z)$.

2. Trinomial that is a perfect square. $x^2 \pm 2xy + y^2 = (x \pm y)^2$.

3. The difference of two squares. (a) Two terms. $x^2 - y^2 = (x + y)(x - y)$. (b) Four terms. $x^2 + 2xy + y^2 - m^2 = (x + y + m)(x + y - m)$. (c) Six terms. $x^2 + 2xy + y^2 - p^2 - 2pq - q^2 = [(x + y) + (p + q)][(x + y) - (p + q)]$. (d) Incomplete square. $x^4 + x^2 y^2 + y^4 = x^4 + 2x^2 y^2 + y^4 - x^2 y^2 = (x^2 + y^2 + xy)(x^2 + y^2 - xy)$.

4. Trinomial of the form $x^2 + bx + c$. $x^2 - 5x + 6 = (x - 2)(x - 3)$.

5. Trinomial of the form $ax^2 + bx + c$. $20x^2 + 7x - 6 = (4x + 3)(5x - 2)$.

A Review of Algebra, by Romeyn Henry Rivenburg

6. Sum or difference of two cubes. See "Special Rules," 7 and 8. two like powers. See "Special Rules," 9.

7. Common polynomial factor. Grouping. $t^2 p + t^2 q - 2mp - 2mq = t^2(p + q) - 2m(p + q) = (p + q)(t^2 - 2m)$.

8. Factor Theorem. $x^3 + 17x - 18 = (x - 1)(x^2 + x + 18)$.

~H. C. F. and L. C. M.~

$a^2 + 2a - 3 = (a + 3)(a - 1)$.

$a^2 + 7a + 12 = (a + 3)(a + 4)$.

$a^4 + 27a = a(a + 3)(a^2 - 3a + 9)$.

H. C. F. $= a + 3$.

L. C. M. $= (a + 3)(a - 1)(a + 4)a(a^2 - 3a + 9)$.

~Fractions~

Reduction to lowest terms.

Reduction of a mixed number to an improper fraction.

Reduction of an improper fraction to a mixed number.

Addition and subtraction of fractions.

Multiplication and division of fractions.

Law of signs in division, changing signs of factors, etc.

Complex fractions.

~Simultaneous Equations~

Solved by addition or subtraction. substitution. comparison.

Graphical representation.

~Involution~

Law of signs.

Binomial theorem laws.

Expansion of monomials and fractions. binomials. trinomials.

~Evolution~

Law of signs.

Evolution of monomials and fractions.

Square root of algebraic expressions.

Square root of arithmetical numbers.

Optional Cube root of algebraic expressions. Cube root of arithmetical numbers.

~Theory of Exponents~

Proofs: $a^m \times a^n = a^{(m+n)}$; $(a^m)/(a^n) = a^{(m-n)}$; $(a^m)^n = a^{(mn)}$; $[a^{(mn)}]^{(1/n)} = a^m$; $(a/b)^n = (a^n)/(b^n)$; $(abc)^n = a^n b^n c^n$.

Meaning of fractional exponent. zero exponent. negative exponent.

Four rules To multiply quantities having the same base, add exponents. To divide quantities having the same base, subtract exponents. To raise to a power, multiply exponents. To extract a root, divide the exponent of the power by the index of the root.

~Radicals~

Radical in its simplest form.

Transformation of radicals Fraction under the radical sign. Reduction to an entire surd. Changing to surds of different order. Reduction to simplest form.

Addition and subtraction of radicals.

Multiplication and division of radicals $a^{1/n} \cdot b^{1/n} = [ab]^{1/n}$. $([ab]^{1/n})/(a^{1/n}) = b^{1/n}$.

Rationalization Monomial denominator. Binomial denominator. Trinomial denominator.

Square root of a binomial surd.

Radical equations. *Always* check results to avoid extraneous roots.

~Quadratic Equations~

Pure. $x^2 = a$.

Affected. $ax^2 + bx + c = 0$.

Methods of solving Completing the square. Formula. Developed from $ax^2 + bx + c = 0$. Factoring.

Equations in the quadratic form.

Properties of quadratics $r1 = -b/2a + ([b^2 - 4ac]^{1/2})/(2a)$. $r2 = -b/2a - ([b^2 - 4ac]^{1/2})/(2a)$. Then $r1 + r2 = -b/a$. $r1 \cdot r2 = c/a$. Discriminant, $b^2 - 4ac$, and its discussion. Nature or character of the roots.

~Simultaneous Quadratics~

CASE I.

One equation linear.

The other quadratic. $3x - y = 12$, $x^2 - y^2 = 16$.

CASE II.

Both equations homogeneous and of the second degree. $x^2 - xy + y^2 = 21$, $y^2 - 2xy = -15$.

CASE III.

Any two of the quantities $x + y$, $x^2 + y^2$, xy, $x^3 + y^3$, $x^3 - y^3$, $x - y$, $x^2 \pm xy + y^2$, etc., given. $x^2 + y^2 = 41$, $x + y = 9$.

CASE IV. Both equations symmetrical or symmetrical except for sign. Usually one equation of high degree, the other of the first degree. $x^5 + y^5 = 242$, $x + y = 2$.

CASE V. Special Devices

I. Solve for a compound unknown, like xy, $x + y$, $1/(xy)$, etc., first. $x^2 y^2 + xy = 6$, $x + 2y = -5$.

II. Divide the equations, member by member. $x^4 - y^4 = 20$, $x^2 - y^2 = 5$.

III. Eliminate the quadratic terms. $4x + 3y = 2xy$, $7x - 5y = 5xy$.

~Ratio and Proportion~

Proportionals mean, third, fourth.

Theorems 1. Product of means equals product of extremes. 2. If the product of two numbers equals the product of two other numbers, either pair, etc. 3. Alternation. 4. Inversion. 5. Composition. 6. Division. 7. Composition and division. 8. In a series of equal ratios, the sum of the antecedents is to the sum of the consequents as any antecedent, etc.

Special method of proving four quantities in proportion. Let $a/b = x$, $a = bx$, etc.

~Progressions~

Development of formulas. $\{\, l = ar^{(n-1)}.\ \{\, l = a + (n-1)d.\ \{\, S = (ar^n - a)/(r-1).\ \{\, S = (n/2)(a + l).\ \{\, S = (rl - a)/(r-1).\ \{\, S = (n/2)[2a + (n-1)d].\ \{\, S[\infty] = (a)/(1-r).$

Insertion of means Arithmetical. Geometrical.

~Binomial Theorem~

Review of binomial theorem laws. See Involution.

Expansion of $(a + b)^n$.

Finding any term by key number method. $r^{(th)}$ or $(r + 1)^{(th)}$ term method.

A REVIEW OF ALGEBRA

ORDER OF OPERATIONS, EVALUATION, PARENTHESES

Order of operations: First of all, raising to a power and extracting a root. Next, multiplication and division. Last of all, addition and subtraction.

Find the value of:

1. $5 \cdot 2^2 - 25^{(1/2)} \div 5 + 2^2 \cdot 8 \div 4 - 2.$

2. $(3 \times 6 \div 9)/2 - 2[100^{(1/2)}] \div 5 + 4 \cdot 2^3 - (14 \cdot 2)/28.$

3. $9 \cdot 2 \div 6 + 3 - 2 \cdot 4^2 \div 8^{(1/3)} - 4 + (3 \cdot 2^2)/6.$

Evaluate:

4. $(a^4 - a^3 + b^3)/([a^2 b^2]^{(1/2)}) + (c[a^{(1/2)}] + a^3bc)/(abc)$, if $a = 1$, $b = 2$, $c = 3$.

5. $t^{(1/3)} + [tm]^{(1/3)} + m^{(1/3)}$, if $t = 8$, $m = 27$.

6. $(2[3 + 2d + a]^{1/2})/(3[a + b - cx - c]^{1/2}) + ((3c - d)x)/(7ad - [abc]^{1/2})$, if $a = 5$, $b = 3$, $c = -1$, $d = -2$, $x = 0$.

7. $a - \{5b - [a - (3c - 3b) + 2c - 3(a - 2b - c)]\}$, if $a = -3$, $b = 4$, $c = -5$. (*Yale.*)

Simplify:

8. $m - [2m - \{3r - (4r - 2m)\}]$.

9. $2a - [5d + \{3c - (a + [2d - 3a + 4c])\}]$.

10. $3c^2 + c(2a - [6c - \{3a + c - 4a\}])$.

SPECIAL RULES OF MULTIPLICATION AND DIVISION

Give results by inspection:

1. $(g + 1/2\, k)^2$.

2. $(s - (2m)/3)^2$.

3. $(2v + 3w)(2v - 3w)$.

4. $(x + 3ts)(x - 7ts)$.

5. $(2l + 3g)(4l - 11g)$.

6. $(a - (2b)/3 + c - d)^2$.

7. $(x^3 + 8m^3)/(x + 2m)$.

8. $(y^3 - 27k^{3m})/(y - 3k^m)$.

9. $(c^5 - d^5)/(c - d)$.

10. $(e^5 + d^5)/(e + d)$.

11. $(x^4 - y^4)/(x - y)$.

12. $(x^4 - y^4)/(x + y)$.

13. $(a - .03)(a - .0007)$.

14. $(g^n - 1/2)(g^n + 3/4)$.

15. $(t^7 - v^{7/2})/(t - v^{1/2})$.

16. $(k^{32} + 1)(k^{16} + 1)(k^8 + 1)(k^4 + 1)(k^2 + 1)(k + 1)(k - 1)$.

17. $[(a + b) + (c + d)][(a + b) - (c + d)]$.

18. $(p - q + r - s)(p - q - r + s)$.

19. $(3m - n - l + 2r)(3m + n - l - 2r)$.

20. $(x + 5)(x - 2)(x - 5)(x + 2)$.

21. $(a^2 + b^2 - c - 2d + 3e)^2$.

22. $(s + t - 2v/5 + 3w/6 + z^2)^2$.

23. $(x^5 + 32)/(x + 2)$.

~References:~ The chapter on Special Rules of Multiplication and Division in any algebra. Special Rules of Multiplication and Division in the Outline in the front of the book.

CASES IN FACTORING

The number of terms in an expression usually gives the clue to the possible cases under which it may come. By applying the *test* for each and eliminating the *possible* cases one by one, the right case is readily found. Hence, the number of terms in the expression and a ready and accurate knowledge of the Cases in Factoring are the real keys to success in this vitally important part of algebra.

CASE I. A common monomial factor. Applies to any number of terms.

A Review of Algebra, by Romeyn Henry Rivenburg

$5cx - 5ct + 5cv - 15c^2m + 25c^3m^2 = 5c(x - t + v - 3cm + 5c^2m^2)$.

CASE II. A trinomial that is a perfect square. Three terms.

$x^2 \pm 2xm + m^2 = (x \pm m)^2$.

CASE III. The difference of two squares.

A. Two terms. $x^2 - y^2 = (x + y)(x - y)$.

B. Four terms.

$x^2 + 2xy + y^2 - m^2 = (x^2 + 2xy + y^2) - m^2 = (x + y + m)(x + y - m)$

C. Six terms. $x^2 - 2xy + y^2 - m^2 - 2mn - n^2 = (x^2 - 2xy + y^2) - (m^2 + 2mn + n^2) = (x - y)^2 - (m + n)^2 = [(x - y) + (m + n)][(x - y) - (m + n)]$.

D. An incomplete square. Three terms, and 4th powers or multiples of 4.

$c^4 + c^2d^2 + d^4 = c^4 + 2c^2d^2 + d^4 - c^2d^2 = (c^2 + d^2)^2 - c^2d^2 = (c^2 + d^2 + cd)(c^2 + d^2 - cd)$.

CASE IV. A trinomial of the form $x^2 + bx + c$. Three terms.

$x^2 + x - 30 = (x + 6)(x - 5)$.

CASE V. A trinomial of the form $ax^2 + bx + c$. Three terms.

$20x^2 + 7x - 6 = (4x + 3)(5x - 2)$.

CASE VI.

A. The sum or difference of two cubes. Two terms.

$x^3 + y^3 = (x + y)(x^2 - xy + y^2); x^3 - y^3 = (x - y)(x^2 + xy + y^2)$.

B. The sum or difference of two like powers. Two terms.

$x^4 - y^4 = (x - y)(x^3 + x^2y + xy^2 + y^3)$; $x^5 + y^5 = (x + y)(x^4 - x^3y + x^2y^2 - xy^3 + y^4)$.

CASE VII. A common polynomial factor. Any *composite* number of terms.

$t^2 p + t^2 q - t^2 r - g^2 p - g^2 q + g^2 r = t^2 (p + q - r) - g^2 (p + q - r) = (p + q - r)(t^2 - g^2) = (p + q - r)(t + g)(t - g)$.

CASE VIII. The Factor Theorem. Any number of terms.

$x^3 + 17x - 18 = (x - 1)(x^2 + x + 18)$.

FACTORING

Review the *Cases in Factoring* (see Outline on preceding pages) and write out the prime factors of the following:

1. $8a^{13} + am^{12}$.

2. $x^7 + y^7$.

3. $4x^2 + 11x - 3$.

4. $m^2 + n^2 - (1 + 2mn)$.

5. $-x^2 + 2x - 1 + x^4$.

6. $x^{16} - y^{16}$. (Five factors.)

7. $(x + 1)^2 - 5x - 29$.

8. $x^4 + x^2 y^2 + y^4$.

9. $x^4 - 11x^2 + 1$.

10. $x^{2m} + 2 + 1/(x^{2m})$.

11. $x^{6m} + 13x^{3m} + 12$.

12. $4a^2 b^2 - (a^2 + b^2 - c^2)^2$.

13. $(x^2 - x - 6)(x^2 - x - 20)$.

14. $a^4 - 8a - a^3 + 8$.

15. $p^3 + 7p^2 + 14p + 8$.

16. $18a^2 b + 60ab^2 + 50b^3$.

17. $x^3 - 7x + 6$.

18. $24c^2 d^2 - 47cd - 75$.

19. $(a^2 - b^2)^2 - (a^2 - ab)^2$.

20. $a^2 x^3 - (8a^2)/(y^3) - x^3 + 8/(y^3)$.

21. $gt - gk + gl^2 + xt - xk + xl^2$.

22. $(m - n)(2a^2 - 2ab) + (n - m)(2ab - 2b^2)$.

23. $a^2 - x^2 - y^2 + b^2 + 2ab + 2xy$.

24. $(2c^2 + 3d^2)a + (2a^2 + 3c^2)d$.

25. $(n(n - 1))/(1 \cdot 2) \, a^{n-2} b^2 + (n(n - 1)(n - 2))/(1 \cdot 2 \cdot 3) \, a^{n-3} b^3$.

26. $(x - x^2)^3 + (x^2 - 1)^3 + (1 - x)^3$. (*M. I. T.*)

27. $(27y^3)^2 - 2(27y^3)(8b^3) + (8b^3)^2$. (*Princeton.*)

28. $(a^3 + 8b^3)(a + b) - 6ab(a^2 - 2ab + 4b^2)$. (*M. I. T.*)

Solve by factoring:

29. $x^3 = x$.

30. $z^2 - 4z - 45 = 0$.

31. $x^3 - x^2 = 4x - 4$.

~Reference:~ The chapter on Factoring in any algebra.

HIGHEST COMMON FACTOR AND LOWEST COMMON MULTIPLE

Define H. C. F. and L. C. M.

Find by factoring the H. C. F. and L. C. M.:

1. $3x^2 - 3x$, $12x^2 (x^2 - 1)$, $18x^3 (x^3 - 1)$.

2. $(x^2 - 1)(x^2 + 5x + 6)$, $(x^2 + 3x)(x^2 - x - 6)$. (*Harvard.*)

3. $x^2 - y^2$, $x^2 + y^2$, $x^3 + y^3$, $x^6 + y^6$, $x^6 - y^6$. (*College Entrance Board.*)

4. $x^3 + x^2 - 2$, $x^3 + 2x^2 - 3$. (*Cornell.*)

5. $x^5 - 2x^4 + x^2$, $2x^4 - 4x^3 - 4x + 6$. (*Yale.*)

6. $x^2 + a^2 - b^2 + 2ax$, $x^2 - a^2 + b^2 + 2bx$, $x^2 - a^2 - b^2 - 2ab$. (*Harvard.*)

7. $2x^2 - x - 15$, $3x^2 - 11x + 6$, $2x^3 - x^2 - 13x - 6$. (*College Entrance Board.*)

8. $(tv - v^2)^3$, $v^3 - t^2v$, $t^3 - v^3$, $v^2 - 2vt + t^2$.

Pick out the H. C. F. and the L. C. M. of the following:

9. $8(x^2 + y)^{(32)} (t^2 + z)^{(19)} (m - n^3)^{(14)}$, $12(x^2 + y)^{(23)} (t^2 + z)^{(41)} (m - n^3)^{(17)}$, $18(m - n^3)^{(11)} (x^2 + y)^{(39)} (t^2 + z)^{(37)}$.

10. $17ax^3(y+z)^{10}(y-x)^{19}(x+z)^{27}$, $34a^2x^4(y+z)^{11}(y-x)^{21}(x+z)^{13}$, $51a^3x^5(y+z)^4(x+z)^{32}(y-x)^{29}$.

~Reference:~ The chapter on H. C. F. and L. C. M. in any algebra.

FRACTIONS

Define: fraction, terms of a fraction, reciprocal of a number.

Look up *the law of signs* as it applies to fractions. Except for this, fractions in algebra are treated exactly the same as they are in arithmetic.

1. Reduce to lowest terms:

(*a*) $32/24$;

(*b*) $(a^6 - x^6)/(a^4 - x^4)$;

(*c*) $[(a+b)^2 - (c+d)^2]/[(a+c)^2 - (b+d)^2]$. (*M. I. T.*)

2. Reduce to a mixed expression:

(*a*) $756/11$;

(*b*) $(a^3 + b^3)/(a-b)$.

3. Reduce to an improper fraction:

(*a*) $45\text{-}1/8$;

(*b*) $9\text{-}11/12$ qt.;

(*c*) $a^2 - ab + b^2 - (b^3)/(a+b)$.

Add:

4. $5/18 + 7/9 + 11/16 + 5/8$.

5. $5/(1 + 2x) - (3x)/(1 - 2x) + (4 - 13x)/(4x^2 - 1)$.

6. $1/[x(x - a)(x - b)] + 1/[a(a - x)(a - b)] + 1/[b(b - x)(b - a)]$.

Multiply:

7. $72/121 \times 55/56 \times 77/90$.

8. $(b - y)/(a^3 + y^3) \times (ca + cy)/(b^2 + by) \times (b^6 + y^6)/(b^2 + y^2) \times b/c$.

Divide:

9. $(12/25) \div (6/50)$.

10. $[1 - (ab)/(a^2 - ab + b^2)] [1 - (ab)/(a^2 + 2ab + b^2)] \div (a^3 - b^3)/(a^3 + b^3)$. (*Yale.*)

11. $[(x^4 - y^4)/(x^2 - y^2) \div (x + y)/(x^2 - xy)] \div [(x^2 + y^2)/(x - y) \div (x + y)/(xy - y^2)]$. (*Sheffield.*)

Simplify:

12. $[(4y)/x - (15y^2)/(x^2) + 4] \div [4 - (16y)/x + (15y^2)/(x^2)] \times [3 - (4x + 20y)/(2x + 5y)]$.

~Reference:~ The chapter on Fractions in any algebra.

COMPLEX FRACTIONS AND FRACTIONAL EQUATIONS

Define a complex fraction.

Simplify:

1. $(3/7 + 4/5)/(2 - 3/7 \cdot 4/5)$.

2. $(2 - 3/2 + 2/3)/(5 - 2/3 + 3/2)$.

3. $2 - 2/(1 - 1/[1 - 1/(1 + 1/2)])$.

4. $a/(b^2) - a/[b^2 + (cb)/(a - c/b)]$. (*Harvard.*)

5. If $m = 1/(a + 1)$, $n = 2/(a + 2)$, $p = 3/(a + 3)$, what is the value of $m/(1 - m) + n/(1 - n) + p/(1 - p)$? (*Univ. of Penn.*)

6. Simplify the expression $\{x + y - 1/[x + y - xy/(x + y)]\}(x^3 - y^3)/(x^2 - y^2)$. (*Cornell.*)

7. Simplify $[1 - (2xy)/((x + y)^2)]/[1 + (2xy)/((x - y)^2)] \div \{(1 - y/x)/(1 + y/x)\}^2$.

8. Solve $(7y + 9)/4 - [y - (2y - 1)/9] = 7$.

9. Solve $2 - 1/3 - (2/5)(x^2 + 3) = (10x)/3 + 1 - (2x^2)/5$.

10. How much water must be added to 80 pounds of a 5 per cent salt solution to obtain a 4 per cent solution? (*Yale.*)

~Reference:~ See Complex Fractions, and the first part of the chapter on Fractional Equations in any algebra.

FRACTIONAL EQUATIONS

1. Solve for each letter in turn $1/b = 1/p + 1/q$.

2. Solve and check:

$(5x + 2)/3 - (3 - (3x - 1)/2) = (3x + 19)/2 - ((x + 1)/6 + 3)$.

3. Solve and check:

$(1/2)(x - a/3) - (1/3)(x - a/4) + (1/4)(x - a/5) = 0$.

4. Solve (after looking up the special *short* method):

$(3x - 1)/30 + (4x - 7)/15 = x/4 - (2x - 3)/(12x - 11) + (7x - 15)/60$.

5. Solve by the special *short* method:

$1/(x - 2) - 1/(x - 3) = 1/(x - 4) - 1/(x - 5)$.

6. At what time between 8 and 9 o'clock are the hands of a watch (*a*) opposite each other? (*b*) at right angles? (*c*) together?

Work out (*a*) and state the equations for (*b*) and (*c*).

7. The formula for converting a temperature of F degrees Fahrenheit into its equivalent temperature of C degrees Centigrade is $C = (5/9)(F - 32)$. Express F in terms of C, and compute F for the values $C = 30$ and $C = 28$. (*College Entrance Exam. Board.*)

8. What is the price of eggs when 2 less for 24 cents raises the price 2 cents a dozen? (*Yale.*)

9. Solve $2/(x - 2) + 2/(4 - x^2) = 5/(x + 2)$.

~Reference:~ The Chapter on Fractional Equations in any algebra. Note particularly the special *short* methods, usually given about the middle of the chapter.

SIMULTANEOUS EQUATIONS

NOTE. Up to this point each topic presented has reviewed to some extent the preceding topics. For example, factoring reviews the special rules of multiplication and division; H. C. F. and L. C. M. review factoring; addition and subtraction of fractions and fractional equations review H. C. F. and L. C. M., etc. From this point on, however, the interdependence is not so marked, and miscellaneous examples illustrating the work already covered will be given very frequently in order to keep the whole subject fresh in mind.

1. Solve by three methods--addition and subtraction, substitution, and comparison: { $5x + y = 11$, { $3x + 2y = 1$.

Solve and check:

2. { $12R_1 - 11R_2 = b + 12c$, { $R_1 + R_2 = 2b + c$.

3. $\{(r-s)/2 = 25/6 - (r+s)/3, \{(r+s-9)/2 - (s-r-6)/3 = 0.$

4. One half of A's marbles exceeds one half of B's and C's together by 2; twice B's marbles falls short of A's and C's together by 16; if C had four more marbles, he would have one fourth as many as A and B together. How many has each? (*College Entrance Board.*)

5. The sides of a triangle are a, b, c. Calculate the radii of the three circles having the vertices as centers, each being tangent externally to the other two. (*Harvard.*)

6. Solve $\{2x + 3y = 7, x - y = 1\}$ graphically; then solve algebraically and compare results. (Use coördinate or squared paper.)

Factor:

7. $x^4 + 4$.

8. $2d^{10} - 1024d$.

9. $2(x^3 - 1) - 7(x^2 - 1)$.

~References:~ The chapters on Simultaneous Equations and Graphs in any algebra.

SIMULTANEOUS EQUATIONS AND INVOLUTION

1. Solve $(3/4)x - (5/3)y = 11\text{-}1/2$, $(5/8)x - (3/2)y = 10\text{-}1/4$.

Look up the method of solving when the unknowns are in the denominator. Should you clear of fractions?

2. Solve $1/x - 1/y - 1/z = 1/a$, $1/y - 1/z - 1/x = 1/b$, $1/z - 1/x - 1/y = 1/c$.

3. Solve graphically and algebraically $2x - y = 4$, $2x + 3y = 12$.

4. Solve graphically and algebraically $3x + 7y = 5$, $8x + 3y = -18$.

Review:

5. The squares of the numbers from 1 to 25.

6. The cubes of the numbers from 1 to 12.

7. The fourth powers of the numbers from 1 to 5.

8. The fifth powers of the numbers from 1 to 3.

9. The binomial theorem laws. (See Involution.)

Expand: (Indicate first, then reduce.)

10. $(b + y)^7$.

11. $[(2a)/3 - 1]^5$.

12. $(x^2 + 2a)^5$.

13. $(x - y + 2z)^3$.

14. A train lost one sixth of its passengers at the first stop, 25 at the second stop, 20% of the remainder at the third stop, three quarters of the remainder at the fourth stop; 25 remain. What was the original number? (*M. I. T.*)

~References:~ The chapter on Involution in any algebra. Also the references on the preceding page.

SQUARE ROOT

Find the square root of:

1. $1 + 16m^6 - 40m^4 + 10m - 8m^3 + 25m^2$.

2. $(a^2)/(x^2) + (6a)/x + 11 + (6x)/a + (x^2)/(a^2)$.

3. Find the square root to three terms of $x^2 + 5$.

4. Find the square root of 337,561.

5. Find the square root of 1823.29.

6. Find to four decimal places the square root of 1.672. (*Princeton.*)

7. Add $2/[(x-1)^3] + 1/[(1-x)^2] - 2/(1-x) - 1/x$.

8. Find the value of: $(64^{1/3} \cdot 12)/24 \div 2 \times 3 - (2 \cdot 7^2)/(14) \div 7 \times 1 + (1^{1/3} \cdot 1^7)/(1 \cdot 1^2) - 4 \cdot 0$.

9. Simplify $[(x+y)^5 + (x-y)^5][(x+y)^5 - (x-y)^5]$.

10. Solve by the short method: $5/(7-x) - [(2\text{-}1/4)x - 3]/4 - (x+11)/8 + (11x+5)/16 = 0$.

11. It takes 3/4 of a second for a ball to go from the pitcher to the catcher, and 1/2 of a second for the catcher to handle it and get off a throw to second base. It is 90 feet from first base to second, and 130 feet from the catcher's position to second. A runner stealing second has a start of 13 feet when the ball leaves the pitcher's hand, and beats the throw to the base by 1/8 of a second. The next time he tries it, he gets a start of only 3-1/2 feet, and is caught by 6 feet. What is his rate of running, and the velocity of the catcher's throw? (*Cornell.*)

~Reference:~ The chapter on Square Root in any algebra.

THEORY OF EXPONENTS

Review the proofs, for positive integral exponents, of:

I. $a^m \times a^n = a^{m+n}$.

II. $(a^m)/(a^n) = a^{m-n}$.

III. $(a^m)^n = a^{mn}$.

IV. $[a^{mn}]^{1/n} = a^m$.

V. $[a/b]^n = (a^n)/(b^n)$.

VI. $(abc)^n = a^n b^n c^n$.

~To find the meaning of a fractional exponent.~

Assume that Law I holds for *all* exponents.

If so, $a^{2/3} \cdot a^{2/3} \cdot a^{2/3} = a^{6/3} = a^2$.

Hence, $a^{2/3}$ is *one of the three equal factors* (hence the cube root) of a^2.

Therefore $a^{2/3} = [a^2]^{1/3}$.

In the same way,

$a^{4/5} \cdot a^{4/5} \cdot a^{4/5} \cdot a^{4/5} \cdot a^{4/5} = a^{20/5} = a^4$.

Hence, $a^{4/5}$ is *one of the five equal factors* (hence the fifth root) of a^4.

Therefore $a^{4/5} = [a^4]^{1/5}$.

In the same way, in general, $a^{p/q} = [a^p]^{1/q}$.

Hence, *the numerator of a fractional exponent indicates the power, the denominator indicates the root*.

~To find the meaning of a zero exponent.~

Assume that Law II holds for *all* exponents.

If so, $(a^m)/(a^m) = a^{(m-m)} = a^0$. But by division, $(a^m)/(a^m) = 1$.

Therefore $a^0 = 1$. Axiom I.

~To find the meaning of a negative exponent.~

Assume that Law I holds for *all* exponents.

If so, $a^m \times a^{-m} = a^{(m-m)} = a^0 = 1$.

Hence, $a^m \times a^{-m} = 1$.

Therefore $a^{-m} = 1/(a^m)$.

Rules:

To multiply quantities having the same base, add exponents.

To divide quantities having the same base, subtract exponents.

To raise a quantity to a power, multiply exponents.

To extract a root, divide the exponent of the power by the index of the root.

1. Find the value of $3^2 - 5 \times 4^0 + 8^{-2/3} + 1^{2/5}$.

2. Find the value of $8^{-2/3} + 9^{3/2} - 2^{-2} + 1^{-2/5} - 7^0$.

Give the value of each of the following:

3. $(3^0)/5$, $3/(5^0)$, $(3^0)/(5^0)$, $3^0 \times 5$, 3×5^0, $3^0 \times 5^0$, $3^0 + 5^0$, $3^0 - 5^0$.

4. Express 7^0 as some power of 7 divided by itself.

Simplify:

5. $16^{1/3} \cdot 2^{1/2} \cdot 32^{5/6}$. (Change to the same base first.)

6. $[2/(8^{-3})]^{1/5}$.

7. $[(x^n)^{(n+2)}]/[(x^{(n+1)})(x^{(n-1)})]$.

8. $(x + 3x^{2/3} - 2x^{1/3})(3 - 2x^{-1/3} + 4x^{-2/3})$.

9. $[(a^2b)/(c^2d)]^{1/2} \times [(c^3d)/(ab^3)]^{1/3} \times [(a^{1/3}c)/(b^{1/4}d^{5/12})]^2$.

10. $[(a^{-4})/(b^{-2}c)]^{-3/4} \times [(a^{-1}b[c^{-3}])^{1/2})/(ab^{-1})]^{1/2}$.

11. $[([a^2]^{1/3})/([b^{-1}]^{1/4}) \cdot ([c^{-3}]^{1/2})/(a^{1/3}) \cdot (b^{-1/4}a^{1/3})/(c^{-1})]^{-6}$.

~Reference:~ The chapter on Theory of Exponents in any algebra.

Solve for x:

1. $x^{2/3} = 4$.

2. $x^{-3/4} = 8$.

Factor:

3. $x^{2/3} - 9$.

4. $x^{3/5} + 27$.

5. $x^{2a} - y^{-6}$.

6. $a^{1/3} x^{1/2} - 3a^{1/3} + 5x^{1/2} - 15$.

7. Find the H. C. F. and L. C. M. of $a^2 + a^{3/2} b^{1/2} + a^{1/2} b^{3/2} - b^2$, $a^2 - a^{3/2} b^{1/2} - a^{1/2} b^{3/2} - b^2$.

8. Simplify the product of: $(ayx^{-1})^{1/2}$, $(bxy^{-2})^{1/3}$, and $(y^2 a^{-2} b^{-2})^{1/4}$. (*Princeton.*)

9. Find the square root of: $25a^{4/3}b^{-3} - 10a^{2/3}b^{-3/2} - 49 + 10a^{-2/3}b^{3/2} + 25a^{-4/3}b^3$.

10. Simplify $[(2^{n+2})/(4^{-n}) \div (8^n)/(2^3)]^{1/5}$.

11. Find the value of $(7 \cdot 13^0 \div 7)/(21^0) + 3^0 \times (4^0 \cdot 7^0)/[(7a + b)^0] + 8^{-2/3}$.

12. Express as a power of 2: 8^3; 4^5; $4^3 \cdot 8^{2/3} \cdot 16^{3/4}$.

13. Simplify $\left\{\left[\dfrac{x^{a+1}}{x^{1-a}}\right]^a \div \left[\dfrac{x^a}{x^{1-a}}\right]^{a-1}\right\}^{1/(3a-1)}$.

14. Simplify $\left[\dfrac{x^{5/2} y^{4/3}}{z^{-5/4}} \cdot \dfrac{z^4}{x^{-3} y^{-5/3}} \div \dfrac{y^{-2} z^{1/4}}{x^{-1/2}}\right]^{1/5}$.

15. Expand $(a^{1/2} + b^{1/3})^4$, writing the result with fractional exponents.

~Reference:~ The chapter on Theory of Exponents in any algebra.

RADICALS

1. Review all definitions in Radicals, also the methods of transforming and simplifying radicals. When is *a radical in its simplest form*?

2. Simplify (to simplest form): $[2/3]^{1/2}$; $[1/11]^{1/2}$; $[3/5]^{1/3}$; $3[5/6]^{1/2}$; $(2a/b)[(8b^2)/(27a)]^{1/2}$; $[5/(x^n)]^{1/2n}$; $(a+b)^2[(-a^4)/((a+b)^5)]^{1/3}$; $27^{1/2}$; $[54]^{1/3}$; $-5[125^{1/2}]$.

3. Reduce to entire surds: $2[3^{1/2}]$; $2[3^{1/4}]$; $6[2^{1/3}]$; $a[[b^2]^{1/n}]$; $-3[2^{1/3}]$; $3a[[(a+2)/(6a^2)]^{1/3}]$; $(a+2y)[(a-2y)/(a+2y)]^{1/2}$.

4. Reduce to radicals of lower order (or simplify indices): $[a^2]^{1/4}$; $[a^3]^{1/6}$; $[27a^3]^{1/6}$; $[81\,a^4 x^8]^{1/12}$; $[9x^2 y^4 z^{10}]^{1/2n}$.

5. Reduce to radicals of the same degree (order, or index): $7^{1/2}$ and $[11]^{1/3}$; $5^{1/3}$ and $3^{1/4}$; $7^{1/6}$ and $3^{1/2}$; $[x^m]^{1/n}$ and $[x^n]^{1/m}$; $[c^y]^{1/x}$, $[c^z]^{1/y}$, and $[c^x]^{1/z}$.

6. Which is greater, $3^{1/2}$ or $4^{1/3}$? $[23]^{1/3}$ or $2[2^{1/2}]$?

7. Which is greatest, $3^{1/2}$, $5^{1/3}$, or $7^{1/4}$? Give work and arrange in descending order of magnitude.

Collect:

8. $128^{1/2} - 2[50^{1/2}] + 72^{1/2} - 18^{1/2}$.

9. $2[5/3]^{1/2} + (1/6)60^{1/2} + 15^{1/2} + [3/5]^{1/2}$.

10. $[(m-n)^2 a]^{1/2} + [(m+n)^2 a]^{1/2} - [am^2]^{1/2} + [a(n-m)^2]^{1/2} - a^{1/2}$.

11. A and B each shoot thirty arrows at a target. B makes twice as many hits as A, and A makes three times as many misses as B. Find the number of hits and misses of each. (*Univ. of Cal.*)

~Reference:~ The chapter on Radicals in any algebra (first part of the chapter).

The most important principle in Radicals is the following:

$(ab)^{1/n} = a^{1/n} b^{1/n}$.

Hence $[ab]^{1/n} = a^{1/n} \cdot b^{1/n}$.

Or, $a^{1/n} \cdot b^{1/n} = [ab]^{1/n}$.

From this also $([ab]^{1/n})/(a^{1/n}) = b^{1/n}$.

Multiply:

1. $2[4^{1/3}]$ by $3[6^{1/3}]$.

2. $2^{1/2}$ by $3^{1/3}$.

3. $2^{1/4}$ by $4^{1/6}$.

4. $[a + x^{1/2}]^{1/2}$ by $[a - x^{1/2}]^{1/2}$.

5. $2^{1/2} + 3^{1/2} - 5^{1/2}$ by $2^{1/2} - 3^{1/2} + 5^{1/2}$.

6. $-p/2 + ([p^2 - 4q]^{1/2})/2$ by $-p/2 - ([p^2 - 4q]^{1/2})/2$.

Divide:

7. $27^{1/2}$ by $3^{1/2}$.

8. $4[18^{1/2}]$ by $5[32^{1/2}]$.

9. $3[12]^{1/3}$ by $6^{1/2}$.

10. $3^{1/2}$ by $3^{1/4}$.

11. $6[105^{1/2}] + 18[40^{1/2}] - 45[12^{1/2}]$ by $3[15^{1/2}]$. (*Short division.*)

12. $10[18]^{1/3} - 4[60]^{1/3} + 5[100]^{1/3}$ by $3[30]^{1/3}$.

Rationalize the denominator:

13. $2/(3^{1/2})$; $7/(7^{1/2})$; $5/(2[5^{1/2}])$; $3/([a^2]^{1/5})$; $4/([a^3]^{1/7})$.

14. $2/(2^{1/2}) + 3^{1/2})$; $(a^{1/2} + b^{1/2})/(a^{1/2} - b^{1/2})$; $3/(3 - 3^{1/2})$.

15. $[3^{1/2} + 2^{1/2}]/[6^{1/2} + 3^{1/2} - 2^{1/2}]$.

Review the method of finding the square root of a binomial surd. (By inspection preferably.) Then find square root of:

16. $5 + 2[6^{1/2}]$.

17. $17 - 12[2^{1/2}]$.

18. $7 - 33^{1/2}$.

~Reference:~ The chapter on Radicals in any algebra, beginning at Addition and Subtraction of Radicals.

MISCELLANEOUS EXAMPLES, ALGEBRA TO QUADRATICS

Results by inspection, examples 1-10.

Divide:

1. $(x^{5/17} + y^{5/17})/(x^{1/17} + y^{1/17})$.

2. $(x - y)/(x^{1/3} - y^{1/3})$.

3. $(m^2 + n^2)/(m^{2/3} + n^{2/3})$.

4. $(x - y^2)/(x^{1/3} - [y^2]^{1/3})$.

Multiply:

5. $[a^{-3/4} + 2/(m^{1/2})]^2$.

6. $(K^{-2/7} - g^{-11/25})^2$.

7. $(r^{2s} + l^{-3m})(r^{2s} - l^{-3m})$.

8. $[a^{-2} + b^{-3} - 1/(c^2)]^2$.

9. $(3K^x + 4t^{-3})(3K^x - 7t^{-3})$.

10. $(2y^{2/7} - 40K^3)(3y^{2/7} + 55K^3)$.

Factor:

11. $x^{2/3} - 64$.

12. $y^{3/5} + 27$.

13. $b^{3/2} - 8m^{-1}$.

14. $3p - 8p^{1/2} - 35$.

Factor, using radicals instead of exponents:

15. $60 - 7[3b^{1/2}] - 6b$.

16. $15m - 2[[mn]^{1/2}] - 24n$.

17. $a - b$ (factor as difference of two squares).

A Review of Algebra, by Romeyn Henry Rivenburg

18. a - b (factor as difference of two cubes).

19. a - b (factor as difference of two fourth powers).

20. Find the H. C. F. and L. C. M. of $x^2 + xy^{1/2} - 2y$, $2x^2 + 5xy^{1/2} + 2y$, $2x^2 - xy^{1/2} - y$.

21. Solve (short method) $(x - 7)/(x - 8) - (x - 8)/(x - 9) = (x - 4)/(x - 5) - (x - 5)/(x - 6)$.

22. Simplify $(ab/c + bc/a + ca/b)/(a/bc + b/ca + c/ab) \times [((a + b + c)^2)/(ab + bc + ca) - 2]$. (*Princeton.*)

1. Solve for p: $2^{(p - 3)} = 128$.

2. Solve for t: $t^{3/2} = -27$.

3. Find the square root of 8114.4064. What, then, is the square root of .0081144064? of 811440.64? From any of the above can you determine the square root of .081144064?

4. The H. C. F. of two expressions is $a(a - b)$, and their L. C. M. is $a^2b(a + b)(a - b)$. If one expression is $ab(a^2 - b^2)$, what is the other?

5. Solve (short method): $5/(7 - x) - [(2-1/4)x - 3]/4 - (x + 11)/8 + (11x + 5)/16 = 0$.

6. Solve $2/m - 3/n + 10/p = -3$, $4/m + 5/p + 6/n = 15$, $1/m - 1/n + 5/p = -1/2$.

7. Simplify $21[2/3]^{1/2} - 5[4/5]^{1/2} + 6[4-1/6]^{1/2} - 10[3-1/5]^{1/2} + (40/3)[11-1/4]^{1/2}$.

8. Does $[16 \times 25]^{1/2} = 4 \times 5$? Does $[16 + 25]^{1/2} = 4 + 5$?

9. Write the fraction $5/(4 + 2[3^{1/2}])$ with rational denominator, and find its value correct to two decimal places.

A Review of Algebra, by Romeyn Henry Rivenburg   33

10. Simplify $[\{([p + [p^2 - q]^{1/2}]/2)^{1/2} + ([p - [p^2 - q]^{1/2}]/2)^{1/2}\}^2]/[p + q^{1/2}]$. (*Princeton.*)

1. Rationalize the denominator of $\{6^{1/2} + 3^{1/2} - 3[2^{1/2}]\}/\{6^{1/2} - 3^{1/2} + 3[2^{1/2}]\}$. (*Univ. of Cal.*)

2. Simplify $[2^{n+4} - 2(2^n)]/[2(2^{n+3})]$. (*Univ. of Penn.*)

3. Find the value of $[1 + 8^{-x/3}]/[(8x)^{1/2} + 10^{x-2}]$, when $x = 2$. (*Cornell.*)

4. Find the value of x if $x^{6/5} = y^4$, $y^{2/3} = 9$. (*M. I. T.*)

5. A fisherman told a yarn about a fish he had caught. If the fish were half as long as he said it was, it would be 10 inches more than twice as long as it is. If it were 4 inches longer than it is, and he had further exaggerated its length by adding 4 inches, it would be 1/5 as long as he now said it was. How long is the fish, and how long did he first say it was? (*M. I. T.*)

6. The force *P* necessary to lift a weight *W* by means of a certain machine is given by the formula

$P = a + bW$,

where *a* and *b* are constants depending on the amount of friction in the machine. If a force of 7 pounds will raise a weight of 20 pounds, and a force of 13 pounds will raise a weight of 50 pounds, what force is necessary to raise a weight of 40 pounds? (First determine the constants *a* and *b*.) (*Harvard.*)

7. Reduce to the simplest form: $[[4/[2^{n+2}]]^{1/n}$; $[ax(a^{-1})x - ax^{-1})]/[x^{2/3} - a^{2/3}]$.

8. Determine the H. C. F. and L. C. M. of $(xy - y^2)^3$ and $y^3 - x^2y$. (*College Entrance Board.*)

1. Simplify $(a - 8m)/(a^{1/3} - 2m^{1/3}) - 2a^{1/3}m^{1/3}$.

A Review of Algebra, by Romeyn Henry Rivenburg

2. Simplify, writing the result with rational denominator: $([a^{1/2} + (1)/(x^{-1/2})]^2 - [(1)/(a^{-1/2}) - x^{1/2}]^2) / (x + [a^2 + x^2]^{1/2})$. (*M. I. T.*)

3. Find $[7 - 48^{1/2}]^{1/2}$.

4. Expand $([a^3]^{1/2} - [b^5]^{1/2})^5$.

5. Expand and simplify $(1 - 2[3^{1/2}] + 3[2^{1/2}])^2$.

6. Solve the simultaneous equations $x^{-1/2} + 2y^{-1/2} = 7/6$, $2x^{-1/2} - y^{-1/2} = 2/3$. (*Yale.*)

7. Find to three places of decimals the value of $\{[(a + b)^{-1/3}]/[(11a + b^2)^{1/6}] \cdot [(\{a^3 - b^3\})^{-1/2}]/[(a - b)^{1/2}]\}^{1/2}$, when $a = 5$ and $b = 3$. (*Columbia.*)

8. Show that $(10 - 4[5^{1/2}])/(5 + 3[5^{1/2}])$ is the negative of the reciprocal of $(10 + 4[5^{1/2}])/(5 - 3[5^{1/2}])$. (*Columbia.*)

9. Solve and check $\{5\}/\{[3x + 2]^{1/2}\} = [3x + 2]^{1/2} + [3x - 1]^{1/2}$.

10. Assuming that when an apple falls from a tree the distance (S meters) through which it falls in any time (t seconds) is given by the formula $S = (1/2)gt^2$ (where $g = 9.8$), find to two decimal places the time taken by an apple in falling 15 meters. (*College Entrance Board.*)

Excellent practice may be obtained by solving the ordinary formulas used in arithmetic, geometry, and physics *orally, for each letter in turn.*

ARITHMETIC

$p = br$ $i = prt$ $a = p + prt$

GEOMETRY

$K = (1/2) bh$ $K = bh$ $K = (a^2)/4 \; 3^{1/2}$ $K = (1/2) (b + b') h$ $K = [pi] R^2$
$C = 2 [pi] R$ $K = [pi] R L$ $S = 4 [pi] R^2$ $V = [pi] R^2 H$ $V = (1/3) [pi] R^2 H$
$V = (4/3) [pi] R^3$ $S = ([pi] R^2 E)/(180)$ $C/(C') = R/(R')$ $K/(K') =$

$(R^2)/(R'^2)$

PHYSICS

$v = gt$ $s = (1/2)gt^2$ $s = (v^2)/(2g)$ $C = E/R$ $E = (wv^2)/(2g)$ $e = (4PI^3)/(bh^3 m)$ $E = (mv^2)/(2)$ $t = [pi][l/g]^{(1/2)}$ $F = (mV^2)/(r)$ $mh = (mv^2)/(2g)$ $R = gs/(g + s)$ $E = (4n^2 l^2 w)/(g)$ $C = (5/9)(F - 32)$

QUADRATIC EQUATIONS

1. Define a quadratic equation; a pure quadratic; an affected (or complete) quadratic; an equation in the quadratic form.

2. Solve the pure quadratic $(7)/(3S^2) - (11)/(9S^2) = 5/6$.

Review the first (or usual) method of completing the square. Solve by it the following:

3. $x^2 + 10x = 24$.

4. $2x^2 - 5x = 7$.

5. $(x - 1)/2 + 2/(x - 1) = 2\text{-}1/2$.

6. $ax^2 + bx + c = 0$.

Review the solution by factoring. Solve by it the following:

7. $x^2 + 8x + 7 = 0$.

8. $24x^2 = 2x + 15$.

9. $3 = 10x - 3x^2$.

10. $-7 = 6x - x^2$.

Solve, by factoring, these equations, which are not quadratics:

11. $x^4 = 16$.

A Review of Algebra, by Romeyn Henry Rivenburg

12. $x^3 = 8$.

13. $x^3 = x$.

Review the solution by formula. Solve by it the following:

14. $5x^2 - 6x = 8$.

15. $(1/2)(x + 1) - (x/3)(2x - 1) = -12$.

16. $x^2 + 4ax = 12a^2$.

17. $3x^2 = 2rx + 2r^2$.

Solve graphically:

18. $x^2 - 2x - 8 = 0$.

19. $x^2 + x - 2 = 0$.

~Reference:~ The chapter on Quadratic Equations in any algebra (first part of the chapter).

1. Solve by three methods--formula, factoring, and completing the square: $x^2 + 10x = 24$.

Review equations in the quadratic form and solve:

2. $x^4 - 5x^2 = -4$.

3. $2[x^{(-2)}]^{(1/3)} - 3[x^{(-1)}]^{(1/3)} = 2$.

4. $(x + 3)/(x - 3) + 6 = 5[(x + 3)/(x - 3)]^{(1/2)}$. (Let $y = [(x + 3)/(x - 3)]^{(1/2)}$ and substitute.)

5. $3x^2 - 4x + 2[3x^2 - 4x - 6]^{(1/2)} = 21$.

6. $x^2 + 5x - 5 = (6)/(x^2 + 5x)$.

Solve and check:

7. $[x + 7]^{1/2} + [3x - 2]^{1/2} = (4x + 9)/([3x - 2]^{1/2})$.

8. $[x^2 - 5]^{1/2} + 6/[[x^2 - 5]^{1/2}] = 5$.

9. $(10w)/([10w - 9]^{1/2}) - [10w + 2]^{1/2} = 2/([10w - 9]^{1/2})$.

Give results by inspection:

10. $(a^{1/2} + b^{1/2})(a^{1/2} - b^{1/2})$.

11. $([10 + 19^{1/2}]^{1/2})([10 - 19^{1/2}]^{1/2})$.

12. How many gallons each of cream containing 33% butter fat and milk containing 6% butter fat must be mixed to produce 10 gallons of cream containing 25% butter fat?

13. I have $6 in dimes, quarters, and half-dollars, there being 33 coins in all. The number of dimes and quarters together is ten times the number of half-dollars. How many coins of each kind are there? (*College Entrance Board.*)

~Reference:~ The last part of the chapter on Quadratic Equations in any algebra.

THE THEORY OF QUADRATIC EQUATIONS

~I. To find the sum and the product of the roots.~

The general quadratic equation is

$$ax^2 + bx + c = 0. \quad (1)$$

Or, $x^2 + (b/a)x + c/a = 0. \quad (2)$

To derive the formula, we have by transposing

$$x^2 + (b/a)x = -c/a.$$

Completing the square,

$$x^2 + (b/a)x + [b/2a]^2 = (b^2)/(4a^2) - c/a = (b^2 - 4ac)/(4a^2).$$

Extracting square root, $x + b/2a = [\pm[b^2 - 4ac]^{(1/2)}]/(2a)$.

Transposing, $x = -b/2a \pm [[b^2 - 4ac]^{(1/2)}]/(2a)$.

Hence, $x = [-b \pm [b^2 - 4ac]^{(1/2)}]/(2a)$.

These two values of x we call *roots*.

For convenience represent them by *r1* and *r2*.

Hence, $r1 = -b/2a + [[b^2 - 4ac]^{(1/2)}]/(2a)$. $r2 = -b/2a - [[b^2 - 4ac]^{(1/2)}]/(2a)$. --- Adding, $r1 + r2 = -(2b)/(2a) = -b/a$. (3)

Also, $r1 = -b/2a + [[b^2 - 4ac]^{(1/2)}]/(2a)$. $r2 = -b/2a - [[b^2 - 4ac]^{(1/2)}]/(2a)$. --- Multiplying, $r1\ r2 = (b^2)/(4a^2) - (b^2 - 4ac)/(4a^2) = (b^2 - b^2 + 4ac)/(4a^2) = (4ac)/(4a^2) = c/a$. (4)

Hence we have shown that

$r1 + r2 = -b/a$, and $r1\ r2 = c/a$.

Or, referring to equation (2) above, we have the following rule:

When the coefficient of x^2 is unity, the sum of the roots is the coefficient of x with the sign changed; the product of the roots is the independent term.

EXAMPLES:

1. $x^2 - 9x + 21 = 0$. Sum of the roots = 9. Products of the roots = 21.

2. $3x^2 - 7x - 18 = 0$. Sum of the roots = 7/3. Product of the roots = -6.

3. $-21x = 17 - 4x^2$. Sum of the roots = 21/4. Product of the roots = -17/4.

~II. To find the nature or character of the roots.~

As before, $r_1 = -b/2a + [[b^2 - 4ac]^{(1/2)}]/(2a)$, $r_2 = -b/2a - [[b^2 - 4ac]^{(1/2)}]/(2a)$.

The $[b^2 - 4ac]^{(1/2)}$ determines the *nature* or *character* of the roots; hence it is called the *discriminant*.

~If $b^2 - 4ac$ is positive, the roots are real, unequal, and either rational or irrational.~

~If $b^2 - 4ac$ is negative, the roots are imaginary and unequal.~

~If $b^2 - 4ac$ is zero, the roots are real, equal, and rational.~

EXAMPLES:

1. $x^2 - 4x + 2 = 0$.

$[b^2 - 4ac]^{(1/2)} = [16 - 8]^{(1/2)} = 8^{(1/2)}$. Therefore: The roots are real, unequal, and irrational.

2. $x^2 - 4x + 6 = 0$.

$[b^2 - 4ac]^{(1/2)} = [16 - 24]^{(1/2)} = -8^{(1/2)}$. Therefore: The roots are imaginary and unequal.

3. $x^2 - 4x + 4 = 0$.

$[b^2 - 4ac]^{(1/2)} = [16 - 16]^{(1/2)} = 0^{(1/2)}$. Therefore: The roots are real, equal, and rational.

~III. To form the quadratic equation when the roots are given.~

Suppose the roots are 3, -7.

Then, $x = 3$, Or, $x - 3 = 0$, $x = -7$. $x + 7 = 0$. ------------------ Multiplying to get a quadratic, $(x - 3)(x + 7) = 0$.

Or, $x^2 + 4x - 21 = 0$.

Or, use the sum and product idea developed on the preceding page. The coefficient of x^2 must be unity.

Add the roots and change the sign to get the coefficient of x.

Multiply the roots to get the independent term.

Therefore: The equation is $x^2 + 4x - 21 = 0$.

In the same way, if the roots are $[2 + 3^{(1/2)}]/7$, $[2 - 3^{(1/2)}]/7$, the equation is

$x^2 - (4/7)x + 1/49 = 0$.

Find the sum, the product, and the nature or character of the roots of the following:

1. $x^2 - 7x + 12 = 0$.

2. $9x^2 - 6x + 1 = 0$.

3. $x^2 + 2x + 9734 = 0$.

4. $16 + 5/x = 17/(x^2)$.

5. $(x - 8)/(x - 3) = x$.

6. $(x + 7)(x - 6) = 70$.

7. $x^2 - x(2)^{(1/2)} = 3$.

8. $pr^2 + qr + s = 0$.

Form the equations whose roots are:

9. 5, -3.

10. 2/3, 5/3.

11. c + d, c - d.

12. -3, -5.

13. $[2 \pm -3^{1/2}]/5$.

14. $8/3 + (2/3)37^{1/2}$, $8/3 - (2/3)37^{1/2}$.

15. $[-2 \pm -2^{1/2}]/2$.

16. Solve $x^2 - 3x + 4 = 0$. Check by substituting the values of x; then check by finding the sum and the product of the roots. Compare the amount of labor required in each case.

17. Solve $(x - 3)(x + 2)(x^2 + 3x - 4) = 0$.

18. Is $e^{4z} + 2e^{3z} + e^{2z} + 2e^{z} + 2 + e^{-2z}$ a perfect square?

19. Find the square root (short method): $(x^2 - 1)(x^2 - 3x + 2)(x^2 - x - 2)$.

20. Solve $(1.2x - 1.5)/(1.5) + (.4x + 1)/(.2x - .2) = (.4x + 1)/(.5)$.

21. The glass of a mirror is 18 inches by 12 inches, and it has a frame of uniform width whose area is equal to that of the glass. Find the width of the frame.

OUTLINE OF SIMULTANEOUS QUADRATICS

~Simultaneous Quadratics~

CASE I.

One equation linear. The other quadratic. $2x + y = 7$, $x^2 + 2y^2 = 22$.

METHOD: Solve for x as in terms of y, or *vice versa*, in the linear and substitute in the quadratic.

CASE II.

Both equations homogeneous and of the second degree. $x^2 - xy + y^2 = 39$, $2x^2 - 3xy + 2y^2 = 43$.

METHOD: Let $y = vx$, and substitute in both equations.

ALTERNATE METHOD: Solve for x in terms of y in one equation and substitute in the other.

CASE III.

Any two of the quantities $x + y$ $x^2 + y^2$ xy $x - y$ $x^3 + y^3$ $x^3 - y^3$ $x^2 + xy + y^2$ $x^2 - xy + y^2$ given.

$x + y = 5$, $x^2 - xy + y^2 = 7$.

METHOD: Solve for $x + y$ and $x - y$; then add to get x, subtract to get y.

CASE IV.

Both equations symmetrical or symmetrical except for sign. Usually one equation of high degree, the other of the first degree. $x^5 + y^5 = 242$, $x + y = 2$.

METHOD: Let $x = u + v$ and $y = u - v$, and substitute in both equations.

~Special Devices~

I. Consider some compound quantity like xy, $[x - y]^{1/2}$, $[xy]^{1/2}$, x/y, etc., as the unknown, at first. Solve for the compound unknown, and combine the resulting equation with the simpler original equation.

$x^2 y^2 + xy = 6$, $x + 2y = -5$.

II. Divide the equations member by member. Then solve by Case I, II, or III.

$x^3 - y^3 = 152$, $x - y = 2$.

III. Eliminate the quadratic terms. Then solve by Case I, II, or III.

$xy + x = 15$, $xy + y = 16$.

SIMULTANEOUS QUADRATICS

Solve:

1. $x + y = 7$, $x^2 + 4xy = 57$.

2. $2x^2 = 46 + y^2$, $xy + y^2 = 14$.

3. $x^2 + y^2 = 25$, $x + y = 1$.

4. $x^4 + y^4 = 2$, $x - y = 2$.

5. $x^3 + y^3 = 28$, $x + y = 4$.

6. $x^2 y^2 + xy - 12 = 0$, $x + y = 4$.

7. $2xy - x + 2y = 16$, $3xy + 2x - 4y = 10$.

8. $(3x - 2y)(2x - 3y) = 26$, $x + 1 = 2y$.

9. $4x^2 + 3xy + 2y^2 = 18$, $3x^2 + 2xy - y^2 = 3$.

10. $x^5 + y^5 = 242$, $x + y = 2$.

11. $x - y + [x - y]^{1/2} = 6$, $xy = 5$.

12. $4x^2 - x + y = 67$, $3x^2 - 3y = 27$.

13. $x - y - [x - y]^{1/2} = 2$, $x^3 - y^3 = 2044$. (*Yale.*)

14. $x^2 + xy + x = 14$, $y^2 + xy + y = 28$. (*Princeton.*)

15. $x^2 + y^2 = 13$, $y^2 = 4(x - 2)$. Plot the graph of each equation. (*Cornell.*)

16. $x^2 + y^2 = xy + 37$, $x + y = xy - 17$. (*Columbia.*)

In grouping the answers, be sure to associate each value of x with the corresponding value of y.

17. The course of a yacht is 30 miles in length and is in the shape of a right triangle one arm of which is 2 miles longer than the other. What is the distance along each side?

~Reference:~ The chapter on Simultaneous Quadratics in any algebra.

RATIO AND PROPORTION

1. Define ratio, proportion, mean proportional, third proportional, fourth proportional.

2. Find a mean proportional between 4 and 16; 18 and 50; $12m^2n$ and $3mn^2$.

3. Find a third proportional to 4 and 7; 5 and 10; $a^2 - 9$ and $a - 3$.

4. Find a fourth proportional to 2, 5, and 4; 35, 20, and 14.

5. Write out the proofs for the following, stating the theorem in full in each case:

(*a*) The product of the extremes equals etc.

(*b*) If the product of two numbers equals the product of two other numbers, either pair etc.

(*c*) Alternation.

(*d*) Inversion.

(*e*) Composition.

(*f*) Division.

(*g*) Composition and division.

(*h*) In a series of equal ratios, the sum of the antecedents is to the sum of the consequents etc.

(*i*) Like powers or like roots of the terms of a proportion etc.

6. If x : m :: 13 : 7, write all the possible proportions that can be derived from it. [See (5) above.]

7. Given rs = 161m; write the eight proportions that may be derived from it, and quote your authority.

8. (*a*) What theorem allows you to change any proportion into an equation?

(*b*) What theorem allows you to change any equation into a proportion?

9. If xy = rg, what is the ratio of x to g? of y to r? of y to g?

10. Find two numbers such that their sum, difference, and the sum of their squares are in the ratio 5 : 3 : 51. (*Yale.*)

~Reference:~ The chapter on Ratio and Proportion in any algebra.

An easy and powerful method of proving four expressions in proportion is illustrated by the following example:

Given a : b = c : d;

prove that $3a^3 + 5ab^2 : 3a^3 - 5ab^2 = 3c^3 + 5cd^2 : 3c^3 - 5cd^2$.

Let a/b = r. Therefore a = br.

Also $c/d = r$. Therefore $c = dr$.

Substitute the value of a in the first ratio, and c in the second:

Then

$(3a^3 + 5ab^2)/(3a^3 - 5ab^2) = (3b^3r^3 + 5b^3r)/(3b^3r^3 - 5b^3r) = [b^3r(3r^2 + 5)]/[b^3r(3r^2 - 5)] = (3r^2 + 5)/(3r^2 - 5)$.

Also

$(3c^3 + 5cd^2)/(3c^3 - 5cd^2) = (3d^3r^3 + 5d^3r)/(3d^3r^3 - 5d^3r) = [d^3r(3r^2 + 5)]/[d^3r(3r^2 - 5)] = (3r^2 + 5)/(3r^2 - 5)$.

Therefore $(3a^3 + 5ab^2)/(3a^3 - 5ab^2) = (3c^3 + 5cd^2)/(3c^3 - 5cd^2)$.

Axiom 1.

Or, $3a^3 + 5ab^2 : 3a^3 - 5ab^2 = 3c^3 + 5cd^2 : 3c^3 - 5cd^2$.

If $a : b = c : d$, prove:

1. $a^2 + b^2 : a^2 = c^2 + d^2 : c^2$.

2. $a^2 + 3b^2 : a^2 - 3b^2 = c^2 + 3d^2 : c^2 - 3d^2$.

3. $a^2 + 2b^2 : 2b^2 = ac + 2bd : 2bd$.

4. $2a + 3c : 2a - 3c = 8b + 12d : 8b - 12d$.

5. $a^2 - ab + b^2 : (a^3 - b^3)/a = c^2 - cd + d^2 : (c^3 - d^3)/c$.

6. The second of three numbers is a mean proportional between the other two. The third number exceeds the sum of the other two by 20; and the sum of the first and third exceeds three times the second by 4. Find the numbers.

7. Three numbers are proportional to 5, 7, and 9; and their sum is 14. Find the numbers. (*College Entrance Board.*)

8. A triangular field has the sides 15, 18, and 27 rods, respectively. Find the dimensions of a similar field having 4 times the area.

~ARITHMETICAL PROGRESSION~

1. Define an arithmetical progression.

Learn to derive the three formulas in arithmetical progression:

$l = a + (n - 1)d$, $S = (n/2)(a + l)$, $S = (n/2)[2a + (n - 1)d]$.

2. Find the sum of the first 50 odd numbers.

3. In the series 2, 5, 8, $\cdots$, which term is 92?

4. How many terms must be taken from the series 3, 5, 7, $\cdots$, to make a total of 255?

5. Insert 5 arithmetical means between 11 and 32.

6. Insert 9 arithmetical means between 7-1/2 and 30.

7. Find x, if $3 + 2x$, $5 + 6x$, $9 + 5x$ are in A. P.

8. The 7th term of an arithmetical progression is 17, and the 13th term is 59. Find the 4th term.

9. How can you turn an A. P. into an equation?

10. Given $a = -5/3$, $n = 20$, $S = -5/3$, find d and l.

11. Find the sum of the first n odd numbers.

12. An arithmetical progression consists of 21 terms. The sum of the three terms in the middle is 129; the sum of the last three terms is 237. Find the series. (Look up the short method for such problems.) (*Mass.*

Inst. of Technology.)

13. B travels 3 miles the first day, 7 miles the second day, 11 miles the third day, etc. In how many days will B overtake A who started from the same point 8 days in advance and who travels uniformly 15 miles a day?

~Reference:~ The chapter on Arithmetical Progression in any algebra.

~GEOMETRICAL PROGRESSION~

1. Define a geometrical progression.

Learn to derive the four formulas in geometrical progression:

{ I. $l = ar^{(n-1)}$. {II. $S = (ar^n - a)/(r - 1)$.

{III. $S = (rl - a)/(r - 1)$. { IV. $S_{[infinity]} = (a)/(1 - r)$.

2. *How many terms must be taken from the series 9, 18, 36, ··· to make a total of 567?*

3. *In the G. P. 2, 6, 18, ···, which term is 486?*

4. *Find x, if 2x - 4, 5x - 7, 10x + 4 are in geometrical progression.*

5. *How can you turn a G. P. into an equation?*

6. *Insert 4 geometrical means between 4 and 972.*

7. *Insert 6 geometrical means between 5/16 and 5120.*

8. *Given a = -2, n = 5, l = -32; find r and S.*

9. *If the first term of a geometrical progression is 12 and the sum to infinity is 36, find the 4th term.*

10. *If the series 3-1/3, 2-1/2, ··· be an A. P., find the 97th term. If a G. P., find the sum to infinity.*

11. The third term of a geometrical progression is 36; the 6th term is 972. Find the first and second terms.

12. Insert between 6 and 16 two numbers, such that the first three of the four shall be in arithmetical progression, and the last three in geometrical progression.

13. A rubber ball falls from a height of 40 inches and on each rebound rises 40% of the previous height. Find by formula how far it falls on its eighth descent. (Yale.)

~Reference:~ The chapter on Geometrical Progression in any algebra.

~THE BINOMIAL THEOREM~

1. Review the Binomial Theorem laws. (See Involution.)

Expand:

2. $(b - n)^7$.

3. $(x + x^{-1})^5$.

4. $[a/x - x/a]^6$.

5. $[x/2y - [xy]^{(1/2)}]^5$.

6. $(x^2 - x + 2)^3$.

7. $[(2[b^2]^{(1/3)})/(y) + (3[y^{(1/2)}])/(b^3)]^4$.

8. $(a + b)^n = a^n + na^{(n-1)}b + [n(n-1)]/(1\cdot 2)\, a^{(n-2)}b^2 + [n(n-1)(n-2)]/(1\cdot 2\cdot 3)\, a^{(n-3)}b^3 + [n(n-1)(n-2)(n-3)]/(1\cdot 2\cdot 3\cdot 4)\, a^{(n-4)}b^4 + \cdots$.

Show by observation that the formula for the

$(r + 1)$th term $= [n(n-1)(n-2)\cdots(n-r+1)]/[1\cdot 2\cdot 3\cdot 4 \cdots r]\, a^{(n-r)}b^r$.

9. Indicate what the 97th term of $(a + b)^n$ would be.

10. Using the expansion of $(a + b)^n$ in (8), derive a formula for the rth term by observing how each term is made up, then generalizing.

Using either the formula in (8) or (10), whichever you are familiar with, find:

11. The 4th term of $[a + 1/a]^{30}$.

12. The 8th term of $(1 + x[y^{1/2}])^{13}$.

13. The middle term of $(2a^{3/4} - y[a^{1/3}])^{10}$.

14. The term not containing x in $[x^3 - 2/x]^{12}$.

15. The term containing x^{18} in $[x^2 - a/x]^{15}$.

~Reference:~ The chapter on The Binomial Theorem in any algebra.

~MISCELLANEOUS EXAMPLES, QUADRATICS AND BEYOND~

1. Solve the equation $x^2 - 1.6x - .23 = 0$, obtaining the values of the roots correct to three significant figures. (Harvard.)

2. Write the roots of $(x^2 + 2x)(x^2 - 2x - 3)(x^2 - x + 1) = 0$. (Sheffield Scientific School.)

3. Solve $2[2x + 2]^{1/2} + [2x + 1]^{1/2} = (12x + 4)/([8x + 8]^{1/2})$. (Yale.)

4. Solve the equation $V = (H/3)(B + x + [Bx]^{1/2})$ for x, taking $H = 6$, $B = 8$, and $V = 28$; and verify your result. (Harvard.)

5. Solve { $x : y = 2 : 3$, { $x^2 + y^2 = 5(x + y) + 2$.

6. Solve $2x^2 - 4x + 3[x^2 - 2x + 6]^{1/2} = 15$. (Coll. Ent. Board.)

7. Find all values of x and y which satisfy the equations: $\{x^{1/2} + y^{1/2} = 4$, $\{1/[(x+1)^{1/2} - x^{1/2}] - 1/[(x+1)^{1/2} + x^{1/2}] = y$. (Mass. Inst. of Technology.)

8. If α and β represent the roots of $px^2 + qx + r = 0$, find $\alpha + \beta$, $\alpha - \beta$, and $\alpha\beta$ in terms of p, q, and r. (Princeton.)

9. Form the equation whose roots are $2 + 3^{1/2}$ and $2 - (-3)^{1/2}$.

10. Determine, without solving, the character of the roots of $9x^2 - 24x + 16 = 0$. (College Entrance Board.)

11. If $a:b = c:d$, prove that $a+b:c+d = [a^2+b^2]^{1/2} : [c^2+d^2]^{1/2}$. (College Entrance Board.)

12. Given $a:b = c:d$. Prove that $a^2+b^2 : a^3/(a+b) = c^2+d^2 : c^3/(c+d)$. (Sheffield.)

13. The 9th term of an arithmetical progression is $1/6$; the 16th term is $5/2$. Find the first term. (Regents.)

Solve graphically:

1. $x^2 - x - 6 = 0$.

2. $x^2 + 3x - 10 = 0$.

3. Find four numbers in arithmetical progression, such that the sum of the first two is 1, and the sum of the last two is -19.

4. What number added to 2, 20, 9, 34, will make the results proportional?

5. Find the middle term of $[3a^5 + (b^{3/4})/2]^8$.

6. Solve $(x+1)/(3x+2) = (2x-3)/(3x-2) - 1 - 36/(4-9x^2)$. (Princeton.)

7. A strip of carpet one half inch thick and 29-6/7 feet long is rolled on a roller four inches in diameter. Find how many turns there will be, remembering that each turn increases the diameter by one inch, and that the circumference of a circle equals (approximately) 22/7 times the diameter. (Harvard.)

8. The sum of the first three terms of a geometrical progression is 21, and the sum of their squares is 189. What is the first term? (Yale.)

9. Find the geometrical progression whose sum to infinity is 4, and whose second term is 3/4.

10. Solve $4x + 4[3x^2 - 7x + 3]^{(1/2)} = 3x^2 - 3x + 6$.

11. Solve $\{ 2x^2 + 3xy - 5y^2 = 4, \{ 2xy + 3y^2 = -3$.

12. Two hundred stones are placed on the ground 3 feet apart, the first being 3 feet from a basket. If the basket and all the stones are in a straight line, how far does a person travel who starts from the basket and brings the stones to it one by one?

Solve graphically; and check by solving algebraically:

1. $\{ x^2 + y^2 = 25, \{ x + y = 1$.

2. $x^2 - 3x - 18 = 0$.

3. $x^2 + 3x - 10 = 0$.

Determine the value of m for which the roots of the equation will be equal: (HINT: See page 40. To have the roots equal, $b^2 - 4ac$ must equal 0.)

4. $2x^2 - mx + 12\text{-}1/2 = 0$.

5. $(m - 1)x^2 + mx + 2m - 3 = 0$.

6. If $2a + 3b$ is a root of $x^2 - 6bx - 4a^2 + 9b^2 = 0$, find the other root without solving the equation. (Univ. of Penn.)

7. How many times does a common clock strike in 12 hours?

8. Find the sum to infinity of $2/(2^{1/2})$, $1/(2^{1/2})$, $1/(2[2]^{1/2})$, $\cdots$.

9. Solve $[x/2 + 6/x]^2 - 6[x/2 + 6/x] + 8 = 0$.

10. Find the value of the recurring decimal $2.214214\cdots$.

11. A man purchases a $500 piano by paying monthly installments of $10 and interest on the debt. If the yearly rate is 6%, what is the total amount of interest?

12. The arithmetical mean between two numbers is 42-1/2, and their geometrical mean is 42. Find the numbers. (College Entrance Exam. Board.)

13. If the middle term of $[3x - (1)/(2[x^{1/2}])]^4$ is equal to the fourth term of $[2[x^{1/2}] + 1/2x]^7$, find the value of x. (M. I. T.)

~PROBLEMS~

~Linear Equations, One Unknown~

1. A train running 30 miles an hour requires 21 minutes longer to go a certain distance than does a train running 36 miles an hour. How great is the distance? (Cornell.)

2. A man can walk 2-1/2 miles an hour up hill and 3-1/2 miles an hour down hill. He walks 56 miles in 20 hours on a road no part of which is level. How much of it is up hill? (Yale.)

3. A physician having 100 cubic centimeters of a 6% solution of a certain medicine wishes to dilute it to a 3-1/2% solution. How much water must he add? (A 6% solution contains 6% of medicine and 94% of water.) (Case.)

4. A clerk earned $504 in a certain number of months. His salary was increased 25%, and he then earned $450 in two months less time than it had previously taken him to earn $504. What was his original salary

per month? (College Entrance Board.)

5. A person who possesses $15,000 employs a part of the money in building a house. He invests one third of the money which remains at 6%, and the other two thirds at 9%, and from these investments he obtains an annual income of $500. What was the cost of the house? (M. I. T.)

6. Two travelers have together 400 pounds of baggage. One pays $1.20 and the other $1.80 for excess above the weight carried free. If all had belonged to one person, he would have had to pay $4.50. How much baggage is allowed to go free? (Yale.)

7. A man who can row 4-1/3 miles an hour in still water rows downstream and returns. The rate of the current is 2-1/4 miles per hour, and the time required for the trip is 13 hours. How many hours does he require to return?

~Simultaneous Equations, Two and Three Unknowns~

1. A manual training student in making a bookcase finds that the distance from the top of the lowest shelf to the under side of the top shelf is 4 ft. 6 in. He desires to put between these four other shelves of inch boards in such a way that the book space will diminish one inch for each shelf from the bottom to the top. What will be the several spaces between the shelves?

2. A quantity of water, sufficient to fill three jars of different sizes, will fill the smallest jar 4 times, or the largest jar twice with 4 gallons to spare, or the second jar three times with 2 gallons to spare. What is the capacity of each jar? (Case.)

3. A policeman is chasing a pickpocket. When the policeman is 80 yards behind him, the pickpocket turns up an alley; but coming to the end, he finds there is no outlet, turns back, and is caught just as he comes out of the alley. If he had discovered that the alley had no outlet when he had run halfway up and had then turned back, the policeman would have had to pursue the thief 120 yards beyond the alley before catching him. How long is the alley? (Harvard.)

4. A and B together can do a piece of work in 14 days. After they have worked 6 days on it, they are joined by C who works twice as fast as A. The three finish the work in 4 days. How long would it take each man alone to do it? (Columbia.)

5. In a certain mill some of the workmen receive $1.50 a day, others more. The total paid in wages each day is $350. An assessment made by a labor union to raise $200 requires $1.00 from each man receiving $1.50 a day, and half of one day's pay from every man receiving more. How many men receive $1.50 a day? (Harvard.)

6. There are two alloys of silver and copper, of which one contains twice as much copper as silver, and the other three times as much silver as copper. How much must be taken from each to obtain a kilogram of an alloy to contain equal quantities of silver and copper? (M. I. T.)

7. Two automobiles travel toward each other over a distance of 120 miles. A leaves at 9 A.M., 1 hour before B starts to meet him, and they meet at 12:00 M. If each had started at 9:15 A.M., they would have met at 12:00 M. also. Find the rate at which each traveled. (M. I. T.)

~Quadratic Equations~

1. Telegraph poles are set at equal distances apart. In order to have two less to the mile, it will be necessary to set them 20 feet farther apart. Find how far apart they are now. (Yale.)

2. The distance S that a body falls from rest in t seconds is given by the formula $S = 16t^2$. A man drops a stone into a well and hears the splash after 3 seconds. If the velocity of sound in air is 1086 feet a second, what is the depth of the well? (Yale.)

3. It requires 2000 square tiles of a certain size to pave a hall, or 3125 square tiles whose dimensions are one inch less. Find the area of the hall. How many solutions has the equation of this problem? How many has the problem itself? Explain the apparent discrepancy. (Cornell.)

4. A rectangular tract of land, 800 feet long by 600 feet broad, is divided into four rectangular blocks by two streets of equal width running through it at right angles. Find the width of the streets, if together they cover an area of 77,500 square feet. *(M. I. T.)*

5. (a) The height y to which a ball thrown vertically upward with a velocity of 100 feet per second rises in x seconds is given by the formula, $y = 100x - 16x^2$. In how many seconds will the ball rise to a height of 144 feet?

(b) Draw the graph of the equation $y = 100x - 16x^2$. *(College Entrance Board.)*

6. Two launches race over a course of 12 miles. The first steams 7-1/2 miles an hour. The other has a start of 10 minutes, runs over the first half of the course with a certain speed, but increases its speed over the second half of the course by 2 miles per hour, winning the race by a minute. What is the speed of the second launch? Explain the meaning of the negative answer. *(Sheffield Scientific School.)*

7. The circumference of a rear wheel of a certain wagon is 3 feet more than the circumference of a front wheel. The rear wheel performs 100 fewer revolutions than the front wheel in traveling a distance of 6000 feet. How large are the wheels? *(Harvard.)*

8. A man starts from home to catch a train, walking at the rate of 1 yard in 1 second, and arrives 2 minutes late. If he had walked at the rate of 4 yards in 3 seconds, he would have arrived 2-1/2 minutes early. Find the distance from his home to the station. *(College Entrance Board.)*

~Simultaneous Quadratics~

1. Two cubical coal bins together hold 280 cubic feet of coal, and the sum of their lengths is 10 feet. Find the length of each bin.

2. The sum of the radii of two circles is 25 inches, and the difference of their areas is 125[pi] square inches. Find the radii.

3. The area of a right triangle is 150 square feet, and its hypotenuse is 25 feet. Find the arms of the triangle.

4. The combined capacity of two cubical tanks is 637 cubic feet, and the sum of an edge of one and an edge of the other is 13 feet.

(a) Find the length of a diagonal of any face of each cube.

(b) Find the distance from upper left-hand corner to lower right-hand corner in either cube.

5. A and B run a mile. In the first heat A gives B a start of 20 yards and beats him by 30 seconds. In the second heat A gives B a start of 32 seconds and beats him by 9-5/11 yards. Find the rate at which each runs. (Sheffield.)

6. After street improvement it is found that a certain corner rectangular lot has lost 1/10 of its length and 1/15 of its width. Its perimeter has been decreased by 28 feet, and the new area is 3024 square feet. Find the reduced dimensions of the lot. (College Entrance Board.)

7. A man spends $539 for sheep. He keeps 14 of the flock that he buys, and sells the remainder at an advance of $2 per head, gaining $28 by the transaction. How many sheep did he buy, and what was the cost of each? (Yale.)

8. A boat's crew, rowing at half their usual speed, row 3 miles downstream and back again in 2 hours and 40 minutes. At full speed they can go over the same course in 1 hour and 4 minutes. Find the rate of the crew, and the rate of the current in miles per hour. (College Entrance Board.)

9. Find the sides of a rectangle whose area is unchanged if its length is increased by 4 feet and its breadth decreased by 3 feet, but which loses one third of its area if the length is increased by 16 feet and the breadth decreased by 10 feet. (M. I. T.)

COLLEGE ENTRANCE EXAMINATIONS

~UNIVERSITY OF CALIFORNIA~

ELEMENTARY ALGEBRA

1. If $a = 4$, $b = -3$, $c = 2$, and $d = -4$, find the value of: (a) $ab^3 - 3cd^2 + 2(3a - b)(c - 2d)$. (b) $2a^3 - 3b^4 + (4c^3 + d^3)(4c^2 + d^2)$.

2. Reduce to a mixed number: $(3a^4 - 4a^3 - 10a^2 + 41a - 28)/(a^2 - 3a + 4)$.

Simplify:

3. $(a + 2)/(a^2 + 3a - 40) - (b - 2)/(ab - 5b + 3a - 15)$.

4. $[1 - (2 - 3b - 2c)/(a + 2)] \div (a^2 - 4c^2 + 9b^2 + 6ab)/(2a^2 + a - 6)$.

5. A's age 10 years hence will be 4 times what B's age was 11 years ago, and the amount that A's age exceeds B's age is one third of the sum of their ages 8 years ago. Find their present ages.

6. Draw the lines represented by the equations $3x - 2y = 13$ and $2x + 5y = -4$, and find by algebra the coördinates of the point where they intersect.

7. Solve the equations $\{\ bx - ay = b^2 - ab,\ \{\ y - b = 2(x - 2a)$.

8. Solve $(2x + 1)(3x - 2) - (5x - 7)(x - 2) = 41$.

~COLORADO SCHOOL OF MINES~

ELEMENTARY ALGEBRA

1. Solve by factoring: $x^3 + 30x = 11x^2$.

2. Show that $1 - [(a^2 + b^2 - c^2)/(2ab)]^2 = (a + b + c)(a + b - c)(a - b + c)(b + c - a) \div 4a^2b^2$.

3. How many pairs of numbers will satisfy simultaneously the two equations $\{\ 3x + 2y = 7,\ \{\ x + y = 3$?

Show by means of a graph that your answer is correct.

What is meant by eliminating x in the above equations by substitution? by comparison? by subtraction?

4. Find the square root of 223,728.

5. Simplify: (a) $[1/3]^{(1/2)} + [12]^{(1/2)} - [3/4]^{(1/2)}$. (b) $(-[-3[-4]^{(1/2)}]^{(1/2)})^4$.

6. Solve the equation $.03x^2 - 2.23x + 1.1075 = 0$.

7. How far must a boy run in a potato race if there are n potatoes in a straight line at a distance d feet apart, the first being at a distance a feet from the basket?

~COLUMBIA UNIVERSITY~

ELEMENTARY ALGEBRA COMPLETE

TIME: THREE HOURS

Six questions are required; two from Group A, two from Group B, and both questions of Group C. No extra credit will be given for more than six questions.

Group A

1. (a) Resolve the following into their prime factors: (1) $(x^2 - y^2)^2 - y^4$. (2) $10x^2 - 7x - 6$.

(b) Find the H. C. F. and the L. C. M. of $x^3 - 3x^2 + x - 3$, $x^3 - 3x^2 - x + 3$.

2. (a) Simplify $[x/y + y/x - 2]/[1/x + 1/y] + [x/y + y/x + 2]/[1/x - 1/y]$.

(b) If $x : y = (x - z)^2 : (y - z)^2$, prove that z is a mean proportional between x and y.

3. A crew can row 10 miles in 50 minutes downstream, and 12 miles in an hour and a half upstream. Find the rate of the current and of the crew in still water.

Group B

4. (a) Determine the values of k so that the equation $(2 + k)x^2 + 2kx + 1 = 0$ shall have equal roots.

(b) Solve the equations $x^2 - xy + y^2 = 7$, $2x - 3y = 0$.

(c) Plot the following two equations, and find from the graphs the approximate values of their common solutions: $x^2 + y^2 = 25$, $4x^2 + 9y^2 = 144$.

5. Two integers are in the ratio 4 : 5. Increase each by 15, and the difference of their squares is 999. What are the integers?

6. A man has $539 to spend for sheep. He wishes to keep 14 of the flock that he buys, but to sell the remainder at a gain of $2 per head. This he does and gains $28. How many sheep did he buy, and at what price each?

Group C

7. (a) Find the seventh term of $[a + 1/a]^{13}$.

(b) Derive the formula for the sum of n terms of an arithmetic progression.

8. A ball falling from a height of 60 feet rebounds after each fall one third of its last descent. What distance has it passed over when it strikes the ground for the eighth time?

~CORNELL UNIVERSITY~

ELEMENTARY ALGEBRA

A Review of Algebra, by Romeyn Henry Rivenburg 61

1. Find the H. C. F.: $x^4 - y^4$, $x^3 - xy^2 + x^2y - y^3$, $x^4 + 2x^2y^2 - 3y^4$.

2. Solve the following set of equations: $x + y = -1$, $x + 3y + 2z = -4$, $x - y + 4z = 5$.

3. Expand and simplify: $[2x^3 - 1/x]^7$.

4. An automobile goes 80 miles and back in 9 hours. The rate of speed returning was 4 miles per hour faster than the rate going. Find the rate each way.

5. Simplify: $\{[(x + 1)/(x - 1)]^2 - 2 + [(x - 1)/(x + 1)]^2\} / \{[(x + 1)/(x - 1)]^2 - [(x - 1)/(x + 1)]^2\}$.

6. Solve for x: $(2x + 3)/(x - 1) - 6 = 5/(x^2 + 2x - 3)$.

7. A, B, and C, all working together, can do a piece of work in 2-2/3 days. A works twice as fast as C, and A and C together could do the work in 4 days. How long would it take each one of the three to do the work alone?

~CORNELL UNIVERSITY~

INTERMEDIATE ALGEBRA

1. Solve the following set of equations: $x + y = -1$, $2z + 5w = 1$, $x + 3y + 2z = -4$, $x - y + 4z + 4w = 5$.

2. Simplify: (a) $[6 - 20^{(1/2)}]^{(1/2)}$. (b) $[1 + [x^2 + 1]^{(1/2)}]/[1 + [x^2 + 1]^{(1/2)} + x^2]$.

3. Find, and simplify, the 23d term in the expansion of $[(2x^2)/(3) - 3/4]^{(28)}$.

4. The weight of an object varies directly as its distance from the center of the earth when it is below the earth's surface, and inversely as the square of its distance from the center when it is above the surface. If an object weighs 10 pounds at the surface, how far above,

and how far below the surface will it weigh 9 pounds? (The radius of the earth may be taken as 4000 miles.)

5. Solve the following pair of equations for x and y: $x^2 + y^2 = 4$, $x = (1 + 2^{1/2})y - 2$.

6. Find the value of $[1 + 8^{-x/3}]/[(8x)^{1/2} + 10^{(x-2)}]$, when $x = 2$.

7. From a square of pasteboard, 12 inches on a side, square corners are cut, and the sides are turned up to form a rectangular box. If the squares cut out from the corners had been 1 inch larger on a side, the volume of the box would have been increased 28 cubic inches. What is the size of the square corners cut out? (See the figure on the blackboard.)

~HARVARD UNIVERSITY~

ELEMENTARY ALGEBRA

TIME: ONE HOUR AND A HALF

Arrange your work neatly and clearly, beginning each question on a separate page.

1. Simplify the following expression: $[[1/a + 1/(b + c)]/[1/a - 1/(b + c)]][1 + (b^2 + c^2 - a^2)/(2bc)]$.

2. (a) Write the middle term of the expansion of $(a - b)^{14}$ by the binomial theorem.

(b) Find the value of $a^7 b^7$, if $a = x^{2/7} y^{-3/2}$ and $b = (1/2) x^{-1/7} y^{1/2}$, and reduce the result to a form having only positive exponents.

3. Find correct to three significant figures the negative root of the equation $1 - 2/(x + 1) + 4x/\{(x + 1)^2\} = 0$.

4. Prove the rule for finding the sum of n terms of a geometrical progression of which the first term is a and the constant ratio is r.

Find the sum of 8 terms of the progression 5 + 3-1/3 + 2-2/9 + ⋯.

5. A goldsmith has two alloys of gold, the first being 3/4 pure gold, the second 5/12 pure gold. How much of each must he take to produce 100 ounces of an alloy which shall be 2/3 pure gold?

~HARVARD UNIVERSITY~

ELEMENTARY ALGEBRA

TIME: ONE HOUR AND A HALF

1. Solve the simultaneous equations $x + y = a + b$, $(y + b)/(x + a) = a/b$, and verify your results.

2. Solve the equation $x^2 - 1.6x - 0.23 = 0$, obtaining the values of the roots correct to three significant figures.

3. Write out the first four terms of $(a - b)^7$. Find the fourth term of this expansion when $a = [x^{(-1)} y^{(1/2)}]^{(1/3)}$, $b = [9xy^{(-4)}]^{(1/6)}$, expressing the result in terms of a single radical, and without fractional or negative exponents.

4. Reduce the following expression to a polynomial in a and b: $(6a^3 + 7ab^2 + 12b^3)/(3a^2 - 5ab - 4b^2) - 1/[3/19b - (5a + 4b)/(19a^2)]$.

5. The cost of publishing a book consists of two main items: first, the fixed expense of setting up the type; and, second, the running expenses of presswork, binding, etc., which may be assumed to be proportional to the number of copies. A certain book costs 35 cents a copy if 1000 copies are published at one time, but only 19 cents a copy if 5000 copies are published at one time. Find (a) the cost of setting up the type for the book, and (b) the cost of presswork, binding, etc., per thousand copies.

~HARVARD UNIVERSITY~

ELEMENTARY ALGEBRA

A Review of Algebra, by Romeyn Henry Rivenburg

TIME: ONE HOUR AND A HALF

1. Find the highest common factor and the lowest common multiple of the three expressions $a^4 - b^4$; $a^3 + b^3$; $a^3 + 2a^2 b + 2ab^2 + b^3$.

2. Solve the quadratic equation $x^2 - 1.6x + 0.3 = 0$, computing the value of the larger root correct to three significant figures.

3. In the expression $x^2 - 2xy + y^2 - 4[2^{(1/2)}](x + y) + 8$, substitute for x and y the values $x = (u + v + 1)/[2^{(1/2)}]$, $y = (u - v + 1)/[2^{(1/2)}]$, and reduce the resulting expression to its simplest form.

4. State and prove the formula for the sum of the first n terms of a geometric progression in which a is the first term and r the constant ratio.

5. A state legislature is to elect a United States senator, a majority of all the votes cast being necessary for a choice. There are three candidates, A, B, and C, and 100 members vote. On the first ballot A has the largest number of votes, receiving 9 more votes than his nearest competitor, B; but he fails of the necessary majority. On the second ballot C's name is withdrawn, and all the members who voted for C now vote for B, whereupon B is elected by a majority of 2. How many votes were cast for each candidate on the first ballot?

~MASSACHUSETTS INSTITUTE OF TECHNOLOGY~

ALGEBRA A

TIME: ONE HOUR AND THREE QUARTERS

1. Factor the expressions: $x^3 + x^2 = 2x$. $x^3 + x^2 - 4x - 4$.

2. Simplify the expression: $[1 - (b^2)/(a^2)][1 - (ab - b^2)/(a^2)](a^4)/(a^3 + b^3) \cdot (a - b)/(a^2 + b^2)$.

3. Find the value of $x + [1 + x^2]^{(1/2)}$, when $x = (1/2)[[a/b]^{(1/2)} - [b/a]^{(1/2)}]$.

4. Solve the equations: $(7x + 6)/11 + y - 16 = (5x - 13)/2 - (8y - x)/5$, $3(3x + 4) = 10y - 15$.

5. Solve the equations: $A + C = 2$, $-A + B + C + D = 1$, $2A - B + 2C + D = 5$, $B + D = 1$.

6. Two squares are formed with a combined perimeter of 16 inches. One square contains 4 square inches more than the other. Find the area of each.

7. A man walked to a railway station at the rate of 4 miles an hour and traveled by train at the rate of 30 miles an hour, reaching his destination in 20 hours. If he had walked 3 miles an hour and ridden 35 miles an hour, he would have made the journey in 18 hours. Required the total distance traveled.

~MASSACHUSETTS INSTITUTE OF TECHNOLOGY~

ALGEBRA B

TIME: ONE HOUR AND THREE QUARTERS

1. How many terms must be taken in the series 2, 5, 8, 11, ⋯ so that the sum shall be 345?

2. Prove the formula $x = [-b \pm [b^2 - 4ac]^{(1/2)}]/(2a)$ for solving the quadratic equation $ax^2 + bx + c = 0$.

3. Find all values of a for which $\sqrt{a}$ is a root of $x^2 + x + 20 = 2a$, and check your results.

4. Solve $\{x^2 + 3y^2 = 10, x - y = 2,\}$ and sketch the graphs.

5. The sum of two numbers x and y is 5, and the sum of the two middle terms in the expansion of $(x + y)^3$ is equal to the sum of the first and last terms. Find the numbers.

6. Solve $x^4 - 2x^3 + 3x^2 - 2x + 1 = 0$.

(HINT: Divide by x^2 and substitute $x + 1/x = z$.)

7. In anticipation of a holiday a merchant makes an outlay of $50, which will be a total loss in case of rain, but which will bring him a clear profit of $150 above the outlay if the day is pleasant. To insure against loss he takes out an insurance policy against rain for a certain sum of money for which he has to pay a certain percentage. He then finds that whether the day be rainy or pleasant he will make $80 clear. What is the amount of the policy, and what rate did the company charge him?

~MASSACHUSETTS INSTITUTE OF TECHNOLOGY~

ALGEBRA A

TIME: TWO HOURS

1. Simplify $[m + 1/m]^2 + [n + 1/n]^2 + [mn + 1/mn]^2 - [m + 1/m][n + 1/n][mn + 1/mn]$.

2. Find the prime factors of (a) $(x - x^2)^3 + (x^2 - 1)^3 + (1 - x)^3$. (b) $(2x + a - b)^4 - (x - a + b)^4$.

3. (a) Simplify $[(x^q)/(x^r)]^{(q + r)} [(x^r)/(x^p)]^{(r + p)} [\{x^p/\{x^q\}]^{(p + q)}$.

(b) Show that $([[x]^{[1/(n+1)]}]^{(1/n)})/([[x]^{[1/(n+2)]}]^{[1/(n+1)]}) = \{x^{(1/n)} \cdot [x]^{[1/(n+2)]}\}/\{[x^2]^{[1/(n+1)]}\}$.

4. *Define* homogeneous terms.

For what value of n is $x^n y^{(5 - n/2)} + x^{(n + 1)} y^{(2n - 6)}$ a homogeneous binomial?

5. Extract the square root of $x(x - 2^{(1/2)})(x - 8^{(1/2)})(x - 18^{(1/2)}) + 4$.

6. Two vessels contain each a mixture of wine and water. In the first vessel the quantity of wine is to the quantity of water as 1 : 3, and in the second as 3 : 5. What quantity must be taken from each, so as to form a third mixture which shall contain 5 gallons of wine and 9 gallons

of water?

7. Find a quantity such that by adding it to each of the quantities a, b, c, d, we obtain four quantities in proportion.

8. What values must be given to a and b, so that $(3a + 2b + 17)/2$, $(2a - 3b + 25)/3$, $4 - 5a - 13b$ may be equal?

~MOUNT HOLYOKE COLLEGE~

ELEMENTARY ALGEBRA

TIME: TWO HOURS

1. Factor the following expressions:

(a) $a^{3/4} - b^{3/4}$.

(b) $x^2 y^2 z^2 - x^2 z - y^2 z + 1$.

(c) $16(x + y)^4 - (2x - y)^4$.

2. (a) Simplify

$(a^2 + b^2)\{(b^4)/(b^2 - a^2) - a^2\}/\{a/(a + b) + b/(a - b)\}$.

(b) Extract the square root of $x^4 - 2x^3 + 5x^2 - 4x + 4$.

3. Solve the following equations:

(a) $1/x + 1/y = 5$, $1/(x^2) + 1/(y^2) = 13$.

(b) $x^2 - 5x + 2 = 0$.

(c) $[27x + 1]^{1/2} = 2 - 3[3x^{1/2}]$.

4. Simplify:

(a) $7[54]^{1/3} + 256^{1/6} + [432/(-250)]^{1/3}$.

A Review of Algebra, by Romeyn Henry Rivenburg

(b) $1/[(a-b)(b-c)] + 1/[(c-a)(b-a)]$.

(c) Find $[19 - 8[3^{(1/2)}]]^{(1/2)}$.

5. Plot the graphs of the following system, and determine the solution from the point of intersection: $\{x - 2y = 0, \{2x - 3y = 4$.

6. (a) Derive the formula for the solution of $ax^2 + bx + c = 0$.

(b) Determine the value of m for which the roots of $2x^2 + 4x + m = 0$ are (i) equal, (ii) real, (iii) imaginary.

(c) Form the quadratic equation whose roots are $2 + 3^{(1/2)}$ and $2 - 3^{(1/2)}$.

7. A page is to have a margin of 1 inch, and is to contain 35 square inches of printing. How large must the page be, if the length is to exceed the width by 2 inches?

8. (a) In an arithmetical progression the sum of the first six terms is 261, and the sum of the first nine terms is 297. Find the common difference.

(b) Three numbers whose sum is 27 are in arithmetical progression. If 1 is added to the first, 3 to the second, and 11 to the third, the sums will be in geometrical progression. Find the numbers.

(c) Derive the formula for the sum of n terms of a geometrical progression.

9. (a) Expand and simplify $(2a^2 - 3x^3)^7$.

(b) For what value of x will the ratio $7 + x : 12 + x$ be equal to the ratio $5 : 6$?

~UNIVERSITY OF PENNSYLVANIA~

ELEMENTARY ALGEBRA

A Review of Algebra, by Romeyn Henry Rivenburg

TIME: THREE HOURS

1. Simplify: $[(a + x)/(a - x) - (a - x)/(a + x)] \div (4ax)/(a^2 - x^2)$.

2. Find the H. C. F. and L. C. M. of $10ab^2(x^2 - 2ax)$, $15a^3b(x^2 - ax - 2a^2)$, $25b^3(x^2 - a^2)^2$.

3. A grocer buys eggs at 4 for 7¢. He sells 1/4 of them at 5 for 12¢, and the rest at 6 for 11¢, making 27¢ by the transaction. How many eggs does he buy?

4. Solve for t: $(t + 4a + b)/(t + a + b) - (4t - a - 2b)/(t + a - b) = -3$.

5. Find the square root of $a^2 - (3/2)a^{(3/2)} - (3/2)a^{(1/2)} + (41/16)a + 1$.

6. (a) For what values of m will the roots of $2x^2 + 3mx = -2$ be equal?

(b) If $2a + 3b$ is a root of $x^2 - 6bx - 4a^2 + 9b^2 = 0$, find the other root without solving the equation.

7. (a) Solve for x: $[2x - 3a]^{(1/2)} + [3x - 2a]^{(1/2)} = 3[a^{(1/2)}]$.

(b) Solve for m: $1 - (1)/(2 - m) = 1/(m + 2) + (m - 6)/(4 - m^2)$.

8. Solve the system: $x^2 + 2y^2 = 17$; $xy - y^2 = 2$.

9. Two boats leave simultaneously opposite shores of a river 2-1/4 mi. wide and pass each other in 15 min. The faster boat completes the trip 6-3/4 min. before the other reaches the opposite shore. Find the rates of the boats in miles per hour.

10. Write the sixth term of $[x/(2[y^2]^{(1/3)}) - (y^{(1/2)})/x]^9$ without writing the preceding terms.

11. The sum of the 2d and 20th terms of an A. P. is 10, and their product is 23-47/64. What is the sum of sixteen terms?

~PRINCETON UNIVERSITY~

ALGEBRA A

TIME: TWO HOURS

Candidates who are at this time taking both Algebra A and Algebra B may omit from Algebra A questions 4, 5, and 6, and from Algebra B questions 1 (a), 3, and 4.

1. Simplify $(a^3 + a^2b + ab^2)/(a^2 - 3ab - 4b^2) \div \{(a^2 + 6ab - 7b^2)/(a^2 + 8ab - 9b^2) \cdot (a^3 - b^3)/(a^2 - 7ab + 12b^2)\}$.

2. (a) Divide $a^{5/2} + ab^{3/2} + b^{5/2} - 2a^{1/2}b^2 - a^{3/2}b$ by $a^{3/2} - b^{3/2} + a^{1/2}b - ab^{1/2}$.

(b) Simplify $1/(x^{-1} + y^{-1}) \cdot (x^{1/4}y^{1/2})^3 + 1$.

3. Factor: (a) $(x^2 + 3x)^2 - (2x - 6)^2$.

(b) $a^2 + ac - 4b^2 - 2bc$.

4. Solve $1/(x + 1) - 1/(x - 1) - 1/(x - 3) + 1/(x - 5) = 0$.

5. Solve for x and y: $mx + ax = my - by$, $x - y = a + b$.

6. The road from A to B is uphill for 5 mi., level for 4 mi., and then downhill for 6 mi. A man walks from B to A in 4 hr.; later he walks halfway from A to B and back again to A in 3 hr. and 55 min.; and later he walks from A to B in 3 hr. and 52 min. What are his rates of walking uphill, downhill, and on the level, if these do not vary?

ALGEBRA B

1. Solve

(a) $(x + 1)/(x - 2) + (2x + 1)/(x + 1) + (3x + 3)/(1 - x) = 0$.

(b) $[2x + 7]^{1/2} + [3x - 18]^{1/2} - [7x + 1]^{1/2} = 0$.

(c) $6/(x^2 + 2x) = 5 - 2x - x^2$.

2. Solve for x and y, checking one solution in each problem:

(a) $2x + 3y = 1$, $6/x + 1/y = 2$.

(b) $x^2 = x + y$, $y^2 = 3y - x$.

3. A man arranges to pay a debt of $3600 in 40 monthly payments which form an A. P. After paying 30 of them he still owes 1/3 of his debt. What was his first payment?

4. If 4 quantities are in proportion and the second is a mean proportional between the third and fourth, prove that the third will be a mean prop. between the first and second.

5. In the expansion of $[2x + 1/3x]^6$ the ratio of the fourth term to the fifth is 2 : 1. Find x.

6. Two men A and B can together do a piece of work in 12 days; B would need 10 days more than A to do the whole work. How many days would it take A alone to do the work?

ALGEBRA TO QUADRATICS

1. Simplify $(ab^{-2}c^2)^{1/2} \cdot (a^3b^2c^{-3})^{1/3} + [(a^6)/(b)]^{1/3}$.

2. Simplify $a/[(a - b)(a - c)] + b/[(b - c)(b - a)] + c/[(c - a)(c - b)]$.

3. Factor (a) $x^4 - 10x^2 + 9$.

(b) $x^2 + 2xy - a^2 - 2ay$.

(c) $(a + b)^2 + (a + c)^2 - (c + d)^2 - (b + d)^2$.

4. Find H. C. F. of $x^4 - x^3 + 2x^2 + x + 3$ and $(x + 2)(x^3 - 1)$.

5. Solve $x/(x - 2) + (x - 9)/(x - 7) = (x + 1)/(x - 1) + (x - 8)/(x - 6)$.

6. The sum of three numbers is 51; if the first number be divided by the second, the quotient is 2 and the remainder 5; if the second number be

divided by the third, the quotient is 3 and the remainder 2. What are the numbers?

~SMITH COLLEGE~

ELEMENTARY ALGEBRA

1. Factor $e^{2x} - 2 + e^{-2x}$, $x^{12} - 8$, $x^2 - x - y^2 - y$, $18a^2x^2 - 24axy - 10y^2$.

2. Solve $[7 + 4x + 3[2x^2 + 5x + 7]^{1/2}]^{1/2} - 3 = 0$.

3. The second term of a geometrical progression is $3[2^{1/2}]$, and the fifth term is 3/16. Find the first term and the ratio.

4. Solve the following equations and check your results by plotting:

$\{x^2 + y^2 - xy = 7, \{x + y = 4\}$.

5. Solve

$1/(x^3) + 1/(y^3) = 243/8$, $1/x + 1/y = 9/2$.

6. In an arithmetical progression $d = -11$, $n = 13$, $s = 0$. Find a and l.

7. Expand by the binomial theorem and simplify:

$[(2x)/(y^3) - (y^4)/(x^5 [-6]^{1/2})]^5$.

8. The diagonal of a rectangle is 13 ft. long. If each side were longer by 2 ft., the area would be increased by 38 sq. ft. Find the lengths of the sides.

~SMITH COLLEGE~

ELEMENTARY ALGEBRA

1. Find the H. C. F. of $8x^3 - 27$, $32x^5 - 243$, and $6x^3 - 9x^2 + 4x - 6$.

2. Solve:

(a) $(2x + 5)^{-5} + 31(2x + 5)^{-5/2} = 32$.

(b) $(x - 1)^{1/2} + (3x + 1)^{1/2} = 4$.

3. A farmer sold a horse at $75 for which he had paid x dollars. He realized x per cent profit by his sale. Find x.

4. Find the 13th term and the sum of 13 terms of the arithmetical progression

$(2^{1/2} - 1)/2,\ (2^{1/2})/2,\ (1)/[2([2]^{1/2} - 1)],\ \cdots$.

5. The difference between two numbers is 48. Their arithmetical mean exceeds their geometrical mean by 18. Find the numbers.

6. Expand by the binomial theorem and simplify

$[3a^{-2} - a/[-2]^{1/2}]^5$.

7. Solve:

$1/x + 1/y = 3/2,\ 1/(x^2) + 1/(y^2) = 5/4$.

8. Solve the following equations and check the results by finding the intersections of the graphs of the two equations:

$\{x^2 = 4y,\ \{x + 2y = 4$.

~VASSAR COLLEGE~

ELEMENTARY AND INTERMEDIATE ALGEBRA

Answer any six questions.

1. Find the product of

$[1 + 2a/3 - (5a^2)/(6)]$ and $[2 - 3a/4 + (a^2)/(3)]$.

2. Resolve into linear factors:

(a) $4x^2 - 25$;

(b) $6x^2 - x - 12$;

(c) $a^2b^2 + 1 - a^2 - b^2$;

(d) $y^3 + (x - 3)y^2 - (3x - 2)y + 2x$.

3. Reduce to simplest form:

(a) $z/(1/x - 1/y) + y/(1 - y/x) - x/(1 - x/y)$.

(b) $[-(x^3)^{(1/2)}]^{(1/3)} \times (4y^{(-3)})^{(1/2)}$.

4. (a) Divide $x^{(3/2)} - x^{(-3/2)}$ by $x^{(1/2)} - x^{(-1/2)}$.

(b) Find correct to one place of decimals the value of $[5^{(1/2)} + 7^{(1/2)}]/[2 - 3^{(1/2)}]$.

5. (a) If $a/b = c/d$, show that $(a^2 + c^2)/(b^2 + d^2) = ac/bd$.

(b) Two numbers are in the ratio 3 : 4, and if 7 be subtracted from each the remainders are in the ratio 2 : 3. Find the numbers.

6. Solve the equations:

(a) $(x + 1)/(2) - 3/x = x/3 - (5 - x)/(6)$.

(b) $11x^2 - 11\text{-}1/4 = 9x$.

(c) $\{ x^2 - 2y^2 = 71, \{ x + y = 20$.

7. A field could be made into a square by diminishing the length by 10 feet and increasing the breadth by 5 feet, but its area would then be diminished by 210 square feet. Find the length and the breadth of the field.

~VASSAR COLLEGE~

ELEMENTARY AND INTERMEDIATE ALGEBRA

Answer six questions, including No. 5 and No. 7 or 8. Candidates in Intermediate Algebra will answer Nos. 5-9.

1. Find two numbers whose ratio is 3 and such that two sevenths of the larger is 15 more than one half the smaller.

2. Determine the factors of the lowest common multiple of $3x^4 (x^3 - y^3)$, $15 (x^4 - 2x^2y^2 + y^4)$, and $10y (x^4 + x^2y^2 + y^4)$.

3. Find to two decimal places the value of $[4a^{-2/5} + b^0[ab^{-1}]^{1/2}]^{1/2}$, when $a = -32$ and $b = -8$.

4. Solve the equations: $2x + 5y = 85$, $2y + 5z = 103$, $2z + 5x = 57$.

5. Solve any 3 of these equations:

(a) $x^2 + 44 - 15x = 0$.

(b) $2/x - x/5 = x/20 - 223/30$.

(c) $x^2 + 8x - [4x^2 + 32x + 12]^{1/2} = 21$.

(d) $5/(x + 1) + 8/(x - 2) = 12/(40 - 2x)$.

6. The sum of two numbers is 13, and the sum of their cubes is 910. Find the smaller number, correct to the second decimal place.

7. The sum of 9 terms of an arithmetical progression is 46; the sum of the first 5 terms is 25. Find the common difference.

8. Explain the terms, and prove that if four numbers are in proportion, they are in proportion by alternation, by inversion, and by composition. Find x when $(3 + x)/(3 - x) = (40 + x^3)/(40 - x^3)$.

9. Find the value of x in each of these equations:

(a) $7x^{1/4} - 3x^{1/2} = 2$.

(b) $(x^2 + 2)^{5/2} + 3/\{[x^2 + 2]^{1/2}\} = 4x^2 + 8$.

~YALE UNIVERSITY~

ALGEBRA A

TIME: ONE HOUR

Omit one question in Group II and one in Group III. Credit will be given for six questions only.

Group I

1. *Resolve into prime factors:* (a) $6x^2 - 7x - 20$; (b) $(x^2 - 5x)^2 - 2(x^2 - 5x) - 24$; (c) $a^4 + 4a^2 + 16$.

2. *Simplify* $[5 - (a^2 - 19x^2)/(a^2 - 4x^2)] \div [3 - (a - 5x)/(a - 2x)]$.

3. *Solve* $[2(x - 7)]/(x^2 + 3x - 28) + (2 - x)/(4 - x) - (x + 3)/(x + 7) = 0$.

Group II

4. *Simplify* $[2^{1/2} + 2[3^{1/2}]]/[2^{1/2} - 12^{1/2}]$, *and compute the value of the fraction to two decimal places.*

5. *Solve the simultaneous equations* $\{ x^{-1/2} + 2y^{-1/2} = 7/6,$ $\{ 2x^{-1/2} - y^{-1/2} = 2/3$.

Group III

6. *Two numbers are in the ratio of $c : d$. If a be added to the first and subtracted from the second, the results will be in the ratio of $3 : 2$. Find the numbers.*

7. *A dealer has two kinds of coffee, worth 30 and 40 cents per pound. How many pounds of each must be taken to make a mixture of 70 pounds, worth 36 cents per pound?*

8. A, B, and C can do a piece of work in 30 hours. A can do half as much again as B, and B two thirds as much again as C. How long would each require to do the work alone?

~YALE UNIVERSITY~

ALGEBRA B

TIME: ONE HOUR

Omit one question in Group I and one in Group II. Credit will be given for five questions only.

Group I

1. Solve $(x + a)/(x + b) + (x + b)/(x + a) = 5/2$.

2. Solve the simultaneous equations $\{ x^2y^2 + 28xy - 480 = 0, \{ 2x + y = 11$. Arrange the roots in corresponding pairs.

3. Solve $3x^{-3/2} + 20x^{-3/4} = 32$.

Group II

4. In going 7500 yd. a front wheel of a wagon makes 1000 more revolutions than a rear one. If the wheels were each 1 yd. greater in circumference, a front wheel would make 625 more revolutions than a rear one. Find the circumference of each.

5. Two cars of equal speed leave A and B, 20 mi. apart, at different times. Just as the cars pass each other an accident reduces the power and their speed is decreased 10 mi. per hour. One car makes the journey from A to B in 56 min., and the other from B to A in 72 min. What is their common speed?

Group III

6. Write in the simplest form the last three terms of the expansion of $(4a^{3/2} - a^{1/2} x^{1/3})^8$.

7. (a) *Derive the formula for the sum of an A. P.*

(b_) Find the sum to infinity of the series 1, -1/2, 1/4, -1/8, ⋯. Also find the sum of the positive terms.

End of Project Gutenberg's A Review of Algebra, by Romeyn Henry Rivenburg

*** END OF THIS PROJECT GUTENBERG EBOOK A REVIEW OF ALGEBRA ***

***** This file should be named 38536-8.txt or 38536-8.zip ***** This and all associated files of various formats will be found in:
http://www.gutenberg.org/3/8/5/3/38536/

Produced by Peter Vachuska, Alex Buie, Erica Pfister-Altschul and the Online Distributed Proofreading Team at http://www.pgdp.net

Updated editions will replace the previous one--the old editions will be renamed.

Creating the works from public domain print editions means that no one owns a United States copyright in these works, so the Foundation (and you!) can copy and distribute it in the United States without permission and without paying copyright royalties. Special rules, set forth in the General Terms of Use part of this license, apply to copying and distributing Project Gutenberg-tm electronic works to protect the PROJECT GUTENBERG-tm concept and trademark. Project Gutenberg is a registered trademark, and may not be used if you charge for the eBooks, unless you receive specific permission. If you do not charge anything for copies of this eBook, complying with the rules is very easy. You may use this eBook for nearly any purpose such as creation of derivative works, reports, performances and research. They may be modified and printed and given away--you may do practically ANYTHING with public domain eBooks. Redistribution is subject to the trademark license, especially commercial redistribution.

*** START: FULL LICENSE ***

THE FULL PROJECT GUTENBERG LICENSE PLEASE READ THIS BEFORE YOU DISTRIBUTE OR USE THIS WORK

To protect the Project Gutenberg-tm mission of promoting the free distribution of electronic works, by using or distributing this work (or any other work associated in any way with the phrase "Project Gutenberg"), you agree to comply with all the terms of the Full Project Gutenberg-tm License (available with this file or online at http://gutenberg.org/license).

Section 1. General Terms of Use and Redistributing Project Gutenberg-tm electronic works

1.A. By reading or using any part of this Project Gutenberg-tm electronic work, you indicate that you have read, understand, agree to and accept all the terms of this license and intellectual property (trademark/copyright) agreement. If you do not agree to abide by all the terms of this agreement, you must cease using and return or destroy all copies of Project Gutenberg-tm electronic works in your possession. If you paid a fee for obtaining a copy of or access to a Project Gutenberg-tm electronic work and you do not agree to be bound by the terms of this agreement, you may obtain a refund from the person or entity to whom you paid the fee as set forth in paragraph 1.E.8.

1.B. "Project Gutenberg" is a registered trademark. It may only be used on or associated in any way with an electronic work by people who agree to be bound by the terms of this agreement. There are a few things that you can do with most Project Gutenberg-tm electronic works even without complying with the full terms of this agreement. See paragraph 1.C below. There are a lot of things you can do with Project Gutenberg-tm electronic works if you follow the terms of this agreement and help preserve free future access to Project Gutenberg-tm electronic works. See paragraph 1.E below.

1.C. The Project Gutenberg Literary Archive Foundation ("the Foundation" or PGLAF), owns a compilation copyright in the collection of Project Gutenberg-tm electronic works. Nearly all the individual works in the collection are in the public domain in the United States. If an individual work is in the public domain in the United States and you are located in the United States, we do not claim a right to prevent you from copying, distributing, performing, displaying or creating derivative works based on the work as long as all references to Project Gutenberg are removed. Of course, we hope that you will support the Project Gutenberg-tm mission of promoting free access to electronic works by freely sharing Project Gutenberg-tm works in compliance with the terms of this agreement for keeping the Project Gutenberg-tm name associated with the work. You can easily comply with the terms of this agreement by keeping this work in the same format with its attached full Project Gutenberg-tm License when you share it without charge with others.

1.D. The copyright laws of the place where you are located also govern what you can do with this work. Copyright laws in most countries are in a constant state of change. If you are outside the United States, check the laws of your country in addition to the terms of this agreement before downloading, copying, displaying, performing, distributing or creating derivative works based on this work or any other Project Gutenberg-tm work. The Foundation makes no representations concerning the copyright status of any work in any country outside the United States.

1.E. Unless you have removed all references to Project Gutenberg:

1.E.1. The following sentence, with active links to, or other immediate access to, the full Project Gutenberg-tm License must appear prominently whenever any copy of a Project Gutenberg-tm work (any work on which the phrase "Project Gutenberg" appears, or with which the phrase "Project Gutenberg" is associated) is accessed, displayed, performed, viewed, copied or distributed:

This eBook is for the use of anyone anywhere at no cost and with almost no restrictions whatsoever. You may copy it, give it away or re-use it under the terms of the Project Gutenberg License included with this eBook or online at www.gutenberg.org

1.E.2. If an individual Project Gutenberg-tm electronic work is derived from the public domain (does not contain a notice indicating that it is posted with permission of the copyright holder), the work can be copied and distributed to anyone in the United States without paying any fees or charges. If you are redistributing or providing access to a work with the phrase "Project Gutenberg" associated with or appearing on the work, you must comply either with the requirements of paragraphs 1.E.1 through 1.E.7 or obtain permission for the use of the work and the Project Gutenberg-tm trademark as set forth in paragraphs 1.E.8 or 1.E.9.

1.E.3. If an individual Project Gutenberg-tm electronic work is posted with the permission of the copyright holder, your use and distribution must comply with both paragraphs 1.E.1 through 1.E.7 and any additional terms imposed by the copyright holder. Additional terms will be linked to the Project Gutenberg-tm License for all works posted with the permission of the copyright holder found at the beginning of this work.

1.E.4. Do not unlink or detach or remove the full Project Gutenberg-tm License terms from this work, or any files containing a part of this work or any other work associated with Project Gutenberg-tm.

1.E.5. Do not copy, display, perform, distribute or redistribute this electronic work, or any part of this electronic work, without prominently displaying the sentence set forth in paragraph 1.E.1 with active links or immediate access to the full terms of the Project Gutenberg-tm License.

1.E.6. You may convert to and distribute this work in any binary, compressed, marked up, nonproprietary or proprietary form, including any word processing or hypertext form. However, if you provide access to or distribute copies of a Project Gutenberg-tm work in a format other than "Plain Vanilla ASCII" or other format used in the official version posted on the official Project Gutenberg-tm web site (www.gutenberg.org), you must, at no additional cost, fee or expense to the user, provide a copy, a means of exporting a copy, or a means of obtaining a copy upon request, of the work in its original "Plain Vanilla ASCII" or other form. Any alternate format must include the full Project Gutenberg-tm License as specified in paragraph 1.E.1.

1.E.7. Do not charge a fee for access to, viewing, displaying, performing, copying or distributing any Project Gutenberg-tm works unless you comply with paragraph 1.E.8 or 1.E.9.

1.E.8. You may charge a reasonable fee for copies of or providing access to or distributing Project Gutenberg-tm electronic works provided that

- You pay a royalty fee of 20% of the gross profits you derive from the use of Project Gutenberg-tm works calculated using the method you already use to calculate your applicable taxes. The fee is owed to the owner of the Project Gutenberg-tm trademark, but he has agreed to donate royalties under this paragraph to the Project Gutenberg Literary Archive Foundation. Royalty payments must be paid within 60 days following each date on which you prepare (or are legally required to prepare) your periodic tax returns. Royalty payments should be clearly marked as such and sent to the Project Gutenberg Literary Archive Foundation at the address specified in Section 4, "Information about donations to the Project Gutenberg Literary Archive Foundation."

- You provide a full refund of any money paid by a user who notifies you in writing (or by e-mail) within 30 days of receipt that s/he does not agree to the terms of the full Project Gutenberg-tm License. You must require such a user to return or destroy all copies of the works possessed in a physical medium and discontinue all use of and all access to other copies of Project Gutenberg-tm works.

- You provide, in accordance with paragraph 1.F.3, a full refund of any money paid for a work or a replacement copy, if a defect in the electronic work is discovered and reported to you within 90 days of receipt of the work.

- You comply with all other terms of this agreement for free distribution of Project Gutenberg-tm works.

1.E.9. If you wish to charge a fee or distribute a Project Gutenberg-tm electronic work or group of works on different terms than are set forth in this agreement, you must obtain permission in writing from both the Project Gutenberg Literary Archive Foundation and Michael Hart, the owner of the Project Gutenberg-tm trademark. Contact the Foundation as set forth in Section 3 below.

1.F.

1.F.1. Project Gutenberg volunteers and employees expend considerable effort to identify, do copyright research on, transcribe and proofread public domain works in creating the Project Gutenberg-tm collection. Despite these efforts, Project Gutenberg-tm electronic works, and the medium on which they may be stored, may contain "Defects," such as, but not limited to, incomplete, inaccurate or corrupt data, transcription errors, a copyright or other intellectual property infringement, a defective or damaged disk or other medium, a computer virus, or computer codes that damage or cannot be read by your equipment.

1.F.2. LIMITED WARRANTY, DISCLAIMER OF DAMAGES - Except for the "Right of Replacement or Refund" described in paragraph 1.F.3, the Project Gutenberg Literary Archive Foundation, the owner of the Project Gutenberg-tm trademark, and any other party distributing a Project Gutenberg-tm electronic work under this agreement, disclaim all liability to you for damages, costs and expenses, including legal fees. YOU AGREE THAT YOU HAVE NO REMEDIES FOR NEGLIGENCE, STRICT LIABILITY, BREACH OF WARRANTY OR BREACH OF CONTRACT EXCEPT THOSE PROVIDED IN PARAGRAPH 1.F.3. YOU AGREE THAT THE FOUNDATION, THE TRADEMARK OWNER, AND ANY DISTRIBUTOR UNDER THIS AGREEMENT WILL NOT BE LIABLE TO YOU FOR ACTUAL, DIRECT, INDIRECT, CONSEQUENTIAL, PUNITIVE OR INCIDENTAL DAMAGES EVEN IF YOU GIVE NOTICE OF THE POSSIBILITY OF SUCH DAMAGE.

1.F.3. LIMITED RIGHT OF REPLACEMENT OR REFUND - If you discover a defect in this electronic work within 90 days of receiving it, you can receive a refund of the money (if any) you paid for it by sending a written explanation to the person you received the work from. If you received the work on a physical medium, you must return the medium with your written explanation. The person or entity that provided you with the defective work may elect to provide a replacement copy in lieu of a refund. If you received the work electronically, the person or entity providing it to you may choose to give you a second opportunity to receive the work electronically in lieu of a refund. If the second copy is also defective, you may demand a refund in writing without further opportunities to fix the problem.

1.F.4. Except for the limited right of replacement or refund set forth in paragraph 1.F.3, this work is provided to you 'AS-IS' WITH NO OTHER WARRANTIES OF ANY KIND, EXPRESS OR IMPLIED, INCLUDING BUT NOT LIMITED TO WARRANTIES OF MERCHANTIBILITY OR FITNESS FOR ANY PURPOSE.

1.F.5. Some states do not allow disclaimers of certain implied warranties or the exclusion or limitation of certain types of damages. If any disclaimer or limitation set forth in this agreement violates the law of the state applicable to this agreement, the agreement shall be interpreted to make the maximum disclaimer or limitation permitted by the applicable state law. The invalidity or unenforceability of any provision of this agreement shall not void the remaining provisions.

1.F.6. **INDEMNITY**

- You agree to indemnify and hold the Foundation, the trademark owner, any agent or employee of the Foundation, anyone providing copies of Project Gutenberg-tm electronic works in accordance with this agreement, and any volunteers associated with the production, promotion and distribution of Project Gutenberg-tm electronic works, harmless from all liability, costs and expenses, including legal fees, that arise directly or indirectly from any of the following which you do or cause to occur: (a) distribution of this or any Project Gutenberg-tm work, (b) alteration, modification, or additions or deletions to any Project Gutenberg-tm work, and (c) any Defect you cause.

Section 2. Information about the Mission of Project Gutenberg-tm

Project Gutenberg-tm is synonymous with the free distribution of electronic works in formats readable by the widest variety of computers including obsolete, old, middle-aged and new computers. It exists because of the efforts of hundreds of volunteers and donations from people in all walks of life.

Volunteers and financial support to provide volunteers with the assistance they need, are critical to reaching Project Gutenberg-tm's goals and ensuring that the Project Gutenberg-tm collection will remain freely available for generations to come. In 2001, the Project Gutenberg Literary Archive Foundation was created to provide a secure and permanent future for Project Gutenberg-tm and future generations. To learn more about the Project Gutenberg Literary Archive Foundation and how your efforts and donations can help, see Sections 3 and 4 and the Foundation web page at http://www.pglaf.org.

Section 3. Information about the Project Gutenberg Literary Archive Foundation

The Project Gutenberg Literary Archive Foundation is a non profit 501(c)(3) educational corporation organized under the laws of the state of Mississippi and granted tax exempt status by the Internal Revenue Service. The Foundation's EIN or federal tax identification number is 64-6221541. Its 501(c)(3) letter is posted at http://pglaf.org/fundraising. Contributions to the Project Gutenberg Literary Archive Foundation are tax deductible to the full extent permitted by U.S. federal laws and your state's laws.

The Foundation's principal office is located at 4557 Melan Dr. S. Fairbanks, AK, 99712., but its volunteers and employees are scattered throughout numerous locations. Its business office is located at 809 North 1500 West, Salt Lake City, UT 84116, (801) 596-1887, email business@pglaf.org. Email contact links and up to date contact information can be found at the Foundation's web site and official page at http://pglaf.org

For additional contact information: Dr. Gregory B. Newby Chief Executive and Director gbnewby@pglaf.org

Section 4. Information about Donations to the Project Gutenberg Literary Archive Foundation

Project Gutenberg-tm depends upon and cannot survive without wide spread public support and donations to carry out its mission of increasing the number of public domain and licensed works that can be freely distributed in machine readable form accessible by the widest array of equipment including outdated equipment. Many small donations ($1 to $5,000) are particularly important to maintaining tax exempt status with the IRS.

The Foundation is committed to complying with the laws regulating charities and charitable donations in all 50 states of the United States. Compliance requirements are not uniform and it takes a considerable effort, much paperwork and many fees to meet and keep up with these requirements. We do not solicit donations in locations where we have not received written confirmation of compliance. To SEND DONATIONS or determine the status of compliance for any particular state visit http://pglaf.org

While we cannot and do not solicit contributions from states where we have not met the solicitation requirements, we know of no prohibition against accepting unsolicited donations from donors in such states who approach us with offers to donate.

International donations are gratefully accepted, but we cannot make any statements concerning tax treatment of donations received from outside the United States. U.S. laws alone swamp our small staff.

Please check the Project Gutenberg Web pages for current donation methods and addresses. Donations are accepted in a number of other ways including checks, online payments and credit card donations. To donate, please visit: http://pglaf.org/donate

Section 5. General Information About Project Gutenberg-tm electronic works.

Professor Michael S. Hart is the originator of the Project Gutenberg-tm concept of a library of electronic works that could be freely shared with anyone. For thirty years, he produced and distributed Project Gutenberg-tm eBooks with only a loose network of volunteer support.

Project Gutenberg-tm eBooks are often created from several printed editions, all of which are confirmed as Public Domain in the U.S. unless a copyright notice is included. Thus, we do not necessarily keep eBooks in compliance with any particular paper edition.

Most people start at our Web site which has the main PG search facility:

http://www.gutenberg.org

This Web site includes information about Project Gutenberg-tm, including how to make donations to the Project Gutenberg Literary Archive Foundation, how to help produce our new eBooks, and how to subscribe to our email newsletter to hear about new eBooks.

A Review of Algebra, by Romeyn Henry Rivenburg

A free ebook from http://manybooks.net/

www.ingramcontent.com/pod-product-compliance
Lightning Source LLC
Chambersburg PA
CBHW061444180526
45170CB00004B/1544